Ecotoxicology and Chemistry Applications in Environmental Management

Applied Ecology
and Environmental Management

A SERIES

Series Editor
Sven E. Jørgensen
Copenhagen University, Denmark

Ecotoxicology and Chemistry Applications in Environmental Management, *Sven E. Jørgensen*

Ecological Forest Management Handbook, *Guy R. Larocque*

Handbook of Environmental Engineering, *Frank R. Spellman*

Integrated Environmental Management: A Transdisciplinary Approach, *Sven E. Jørgensen, João Carlos Marques, and Søren Nors Nielsen*

Ecological Processes Handbook, *Luca Palmeri, Alberto Barausse, and Sven E. Jørgensen*

Handbook of Inland Aquatic Ecosystem Management, *Sven E. Jørgensen, Jose Galizia Tundisi, and Takako Matsumura Tundisi*

Eco-Cities: A Planning Guide, *Zhifeng Yang*

Sustainable Energy Landscapes: Designing, Planning, and Development, *Sven Stremke and Andy Van Den Dobbelsteen*

Introduction to Systems Ecology, *Sven E. Jørgensen*

Handbook of Ecological Models Used in Ecosystem and Environmental Management, *Sven E. Jørgensen*

Surface Modeling: High Accuracy and High Speed Methods, *Tian-Xiang Yue*

Handbook of Ecological Indicators for Assessment of Ecosystem Health, Second Edition, *Sven E. Jørgensen, Fu-Liu Xu, and Robert Costanza*

ADDITIONAL VOLUMES IN PREPARATION

Ecotoxicology and Chemistry Applications in Environmental Management

Sven Erik Jørgensen

CRC Press
Taylor & Francis Group
Boca Raton London New York

CRC Press is an imprint of the
Taylor & Francis Group, an **informa** business

CRC Press
Taylor & Francis Group
6000 Broken Sound Parkway NW, Suite 300
Boca Raton, FL 33487-2742

First issued in paperback 2019

ISBN-13: 978-1-4987-1652-9 (hbk)
ISBN-13: 978-0-367-87261-8 (pbk)

Library of Congress Cataloging-in-Publication Data

Names: Jorgensen, Sven Erik, 1934- author.
Title: Ecotoxicology and chemistry applications in environmental management /
Sven Erik Jorgensen.
Description: Boca Raton, FL : Taylor & Francis, 2016. | Series: Applied
ecology and environmental management | Includes bibliographical references
and index.
Identifiers: LCCN 2015045684 | ISBN 9781498716529
Subjects: LCSH: Environmental management. | Environmental toxicology. |
Environmental chemistry.
Classification: LCC GE300 .J659 2016 | DDC 577.27--dc23
LC record available at http://lccn.loc.gov/2015045684

Visit the Taylor & Francis Web site at
http://www.taylorandfrancis.com

and the CRC Press Web site at
http://www.crcpress.com

In Memoriam

Professor Sven Eric Jørgensen passed away on 5 March 2016, while this book was going to press. It was untimely and saddening for all of us here at CRC Press; anyone who knows his work and his outstanding contribution to the Environmental Sciences literature will also recognize this loss.

I feel fortunate and blessed to have known Professor Jørgensen personally as a great author, a visionary book series editor, and an exceptional mentor. Endlessly creative, searching, and challenging, the books he has written and edited during his career will remain a testament to his character.

Thank you, Professor Sven Eric Jørgensen, for your numerous prized works that have enabled so many readers and young scientists to explore the meaning of ecology and environmental management with such passion!

With gratitude,
Irma Shagla Britton
Senior Editor for Environmental Science and Engineering

Contents

Preface xi
Author xiii

1. **Application of Ecotoxicology and Environmental Chemistry in Holistic, Integrated Environmental Management** 1
   1.1 The Seven Management Steps 1
   1.2 Toolboxes Available Today to Develop an Ecological–Environmental Diagnosis 3
   1.3 Toolboxes Available Today to Solve Pollution Problems 6
   1.4 Follow the Recovery Process 8
   1.5 Conclusions about Integrated Environmental Management 9
   References 11

2. **Sources of Toxic Substance Pollution** 13
   2.1 Sources of Pollution 13
   2.2 Distribution in Time and Space 18
   References 26

3. **Properties of Toxic Substances** 27
   3.1 Properties Determining Distribution in the Environment 27
   3.2 Properties (Parameters) Determining Ecotoxicological Effects 32
   References 34

4. **Estimation of Ecotoxicological Parameters** 35
   4.1 Introduction 35
   4.2 Application of Correlation Equations 36
   4.3 Biodegradation 37
   4.4 Estimation of Biological Parameters 39
   4.5 Application of EEP 42
   4.6 Conclusions 52
   References 53

5. **Global Element Cycles** 55
   5.1 The Biochemistry of Nature 55
   5.2 Element Cycles and Recycling in Nature 57
   5.3 The Global Element Cycles 60
   References 65

6. **Toxic Substances in the Environment** 67
   6.1 Inorganic Compounds and Their Effects 67
   6.2 Mercury 72

6.3 Lead...76
6.4 Cadmium..84
6.5 Effects of Organic Compounds...87
    6.5.1 Petroleum Hydrocarbons 87
    6.5.2 PCBs and Dioxins 88
    6.5.3 Pesticides...90
    6.5.4 PAHs...91
    6.5.5 Organometallic Compounds.......................................92
    6.5.6 Detergents (and Soaps)..93
    6.5.7 Synthetic Polymers and Xenobiotics Applied in
         the Plastics Industry 93
References 94

7. Calculations of Reactions and Equilibrium............................97
7.1 Introduction 97
7.2 Equilibrium Constant..98
7.3 Activities and Activity Coefficients..................................... 101
7.4 Mixed Equilibrium Constant 102
7.5 Classification of Chemical Processes and Their
    Equilibrium Constants 104
7.6 Many Simultaneous Reactions.. 108
7.7 Henry's Law.. 108
7.8 Adsorption 110
7.9 Double Logarithmic Diagrams Applied on Acid–Base
    Reactions 112
7.10 Molar Fraction, Alkalinity, and Buffer Capacity 116
7.11 Dissolved Carbon Dioxide.. 119
7.12 Precipitation and Dissolution: Solubility of Hydroxides 124
7.13 Solubility of Carbonates in Open Systems 127
7.14 Solubility of Complexes... 128
7.15 Stability of the Solid Phase 129
7.16 Complex Formation 131
7.17 Environmental Importance of Complex Formation.................... 131
7.18 Conditional Constant 133
7.19 Application of Double Logarithmic Diagrams to
    Determine the Conditional Constants for Complex
    Formation 136
7.20 Redox Equilibria: Electron Activity and Nernst Law.................. 139
7.21 pe as Master Variable.. 142
7.22 Examples of Relevant Processes in the Aquatic Environment... 142
7.23 Redox Conditions in Natural Waters 145
7.24 Construction of pe–pH Diagrams 147
7.25 Redox Potential and Complex Formation............................ 150
7.26 Summary and Conclusions 153
References 153

8. **Environmental Risk Assessment** 155
    8.1   Environmental Risk Analysis 155
    8.2   Development of ERA 160
    References 166

9. **Application of Ecological Models in Environmental**
   Chemistry and Ecotoxicology 167
    9.1   Physical and Mathematical Models 167
    9.2   Models as a Management Tool 169
    9.3   Modeling Components 171
    9.4   Recommended Modeling Procedure 173
    9.5   Overview of Available Ecological Models 184
    9.6   Examples of Application of Models in Ecotoxicology and
           Environmental Chemistry 187
           9.6.1   Food Chain or Food Web Dynamic Models 188
           9.6.2   Steady-State Models of Toxic Substance Mass Flows 189
           9.6.3   Dynamic Model of a Toxic Substance in One
                    Trophic Level 189
           9.6.4   Ecotoxicological Models in Population Dynamics 189
           9.6.5   Ecotoxicological Models with Effect Components 191
    9.7   Models as a Strong Management Tool: Problems and
           Possibilities 193
    References 197

10. **Ecological Indicators and Ecosystem Services as**
    Diagnostic Tools in Ecotoxicology and Environmental
    Chemistry 199
    10.1   Role of Ecosystem Health Assessment in
           Environmental Management 199
    10.2   Criteria for the Selection of Ecological Indicators for EHA 202
    10.3   Classification of Ecosystem Health Indicators 204
    10.4   Ecosystem Services 206
           10.4.1   Provisioning Services 206
           10.4.2   Regulating Services 207
           10.4.3   Supporting Services 207
           10.4.4   Cultural Services 207
    10.5   Value of Ecosystem Services 208
    References 213

11. **Application of Environmental Technology in**
    Environmental Management 215
    11.1   Introduction 215
    11.2   Application of Environmental Technology to Solve
           Wastewater Problems 216
    11.3   Abatement of Air Pollution by Environmental Technology 224

11.4   Solution of Solid Waste Problems by
        Environmental Technology 234
        11.4.1 Methods for Treatment of Solid Waste: An Overview 236
                11.4.1.1 Sludge...237
                11.4.1.2 Domestic Garbage 239
                11.4.1.3 Industrial, Mining, and Hospital Waste 243
                11.4.1.4 Agricultural Waste...245
        References 245

12. **Application of Ecotechnology/Ecological Engineering in
    Environmental Management** 247
    12.1   What Is Ecotechnology and Ecological Engineering? 247
    12.2   Classification of Ecological Engineering/Ecotechnology 249
    12.3   Agricultural Waste and Drainage Water 254
    12.4   Wastewater Treatment by Ecotechnology 255
    12.5   Soil Remediation 258
           12.5.1   Removal and Treatment of Contaminated Soil 258
           12.5.2   *In Situ* Treatment of Contaminated Soil 259
    References 261

13. **Application of Cleaner Production in Environmental
    Management** 263
    13.1   Introduction 263
    13.2   Environmental Management Systems 266
    13.3   Environmental Audit, LCI, and LCA 269
    13.4   Cleaner Production/Technology 270
    13.5   Life Cycle Inventory and Life Cycle Assessment 275
    13.6   Green Chemistry...277
    13.7   Reach Reform..279
    References 280

14. **Implementation of Integrated Environmental Management to
    Solve Toxic Substance Pollution Problems**....................................283
    14.1   Introduction 283
    14.2   Recommendations on Application of Integrated
           Environmental Management..285
    14.3   Following the Recovery Process 288
    References 289

15. **Summary and Conclusions**...291
    15.1   Feasibility of Implementing Integrated Environmental
           Management 291
    15.2   Conclusions...292
    References 294

Index 295

Preface

This book follows the basic ideas of the textbook *Integrated Environmental Management—A Transdisciplinary Approach* by S.E. Jørgensen, J.C. Marques, and S. Nors Nielsen. Environmental management should accordingly be based on seven steps that integrate examination and quantification of an environmental problem and the use of ecological diagnostic tools to develop a diagnosis for ecosystem health, followed by the considerations of using a selected combination of all the applicable toolboxes for the solution of the environmental problem. The textbook presented details on all of the seven steps applied on environmental problems in general. The problems associated with discharge of toxic substances are, however, particularly difficult and complex to solve because there are more than 100,000 different chemical compounds used in a modern industrialized society. It is then implicit that knowledge regarding the important environmental properties of all these chemicals should be available to accomplish the seven steps properly. However, we do not have this enormous knowledge yet, and so we are therefore forced to find some shortcuts to ecotoxicological problems. This book presents in detail how, under these constraints, we could still use the seven steps and set up integrated, holistic environmental management for ecotoxicological problems. This approach would require extensive use of ecotoxicology and environmental chemistry, which are the core topics of the book. Therefore, the book has chapters covering environmental chemical calculations, the QSAR estimation methods, and how toxic substances can interfere with other general environmental problems. There are also chapters focusing on the use of diagnostic ecological subdisciplines and the applicable toolboxes for the solution of environmental problems (environmental technology, ecotechnology, cleaner technology, and environmental legislation). It is my hope that the two books will supplement each other and this book is able to demonstrate that, in spite of the complexity of the seven steps, when the problem is discharge of a toxic substance, integrated environmental management is still applicable.

Additional material is available on the CRC Press website: http://www.crcpress.com/product/isbn/9781498716529.

Sven Erik Jørgensen
Copenhagen

Author

Sven Erik Jørgensen is a professor of environmental chemistry at Copenhagen University. He earned a doctorate in engineering in environmental technology and a doctorate of science in ecological modeling. He is an honorable doctor of science at Coimbra University, Portugal, and at Dar es Salaam University, Tanzania. He was editor-in-chief of *Ecological Modelling* since the journal's inception in 1975 to 2009. He has also been the editor-in-chief of the *Encyclopedia of Ecology*. He is the president of the International Society of Ecological Modelling (ISEM) and was chairman of the International Lake Environment Committee (ILEC) from 1994 to 2006. In 2004, Dr. Jørgensen was awarded the prestigious Stockholm Water Prize and the Prigogine Prize. He was awarded the Einstein Professorship by the Chinese Academy of Science in 2005. In 2007, he received the Pascal medal and was elected a member of the European Academy of Science. He has written more than 360 papers, most of which have been published in international peer-reviewed journals. He has edited or written 70 books. Dr. Jørgensen has given lectures and courses in ecological modeling, ecosystem theory, and ecological engineering worldwide.

1

Application of Ecotoxicology and Environmental Chemistry in Holistic, Integrated Environmental Management

1.1 The Seven Management Steps

Integrated ecological and environmental management means that the environmental problems are viewed from a holistic angle considering the ecosystem as an entity and considering the entire spectrum of diagnostic tools and solutions, including all possible combinations of proposed solutions. Additionally, the experience gained from environmental management in the last 40 years has clearly shown that it is important not just to consider solutions of single problems but to consider *all* the problems associated with a particular ecosystem simultaneously and evaluate *all* the solution possibilities proposed by the relevant disciplines at the same time, or expressed differently, to observe the entire forest and not just single trees. The experience has clearly underlined that there is no alternative to an *integrated* management, at least not on a long-term basis. Fortunately, as presented in this chapter, new ecological subdisciplines have emerged, and they offer toolboxes to perform an integrated ecological and environmental management.

Present-day integrated ecological and environmental management consists of a procedure with the following seven steps (Jørgensen and Nielsen, 2012; Jørgensen et al., 2015):

1. Define the problem.
2. Determine the ecosystems involved.
3. Find and quantify all the sources of the problem.
4. Set up a diagnosis to understand the relation between the problem and the sources.
5. Determine all the tools needed to implement and solve the problem.
6. Implement and integrate the selected solutions.
7. Follow the recovery process.

The same seven steps should be followed when the pollution problem is a toxic substance, which is the focus of this book. When an environmental, ecotoxicological problem has been detected, it is necessary to determine and quantify the problem and all the sources of the problem. It requires the use of analytical methods or a monitoring program. To solve the problem, a clear and unambiguous diagnosis has to be developed: What are the actual problems ecosystems are facing due to pollution by the toxic substance and what are the relationships between the sources and their quantities and the determined problems? Or, expressed differently, to what extent do we solve the problems by reducing or eliminating the different sources of toxic substances discharged into the ecosystem? A holistic integrated approach is needed in most cases because the problems and the corresponding ecological changes or damages in the ecosystems are most often very complex, particularly when several environmental problems are interacting, including the discharge of two or more toxic substances into the same ecosystem. When the first green wave started in the mid-1960s, the tools to answer these questions, which we consider today as very obvious questions in an environmental management context, were not yet developed. We could carry out the first three points of the above shown list but had to stop at point 4 and could at that time only recommend eliminating the source completely or close to completely by the methods available at that time—meaning environmental technology at a lower level than today. Simultaneously, there were no surveys of the ecotoxicological problems and even not a survey of which toxic substances were causing which types of problems in the ecosystems.

Owing to the development of several new ecological subdisciplines, including ecotoxicology, which was hardly defined as a new ecological subdiscipline 45–50 years ago, it is possible today to accomplish points 4–6. The toolboxes we can apply today to carry out point 4 are presented briefly in the next section and in more detail in Chapters 8 through 10. They are the result of the emergence of three new ecological subdisciplines. For a better diagnosis, the following subdisciplines, namely, ecological modeling, ecological indicators, and ecological services, which are covered in Chapters 8 through 10 for toxic substances, are useful. For points 5 and 6, more tools to solve the problems have been developed today—more advanced environmental technology than we had previously, ecological engineering (also denoted as ecotechnology), and cleaner production, which are presented in Chapters 11 through 13. Environmental legislation in this context could also be mentioned as a tool to abate toxic substance pollution, but for this tool, one can refer to Jørgensen et al. (2015), and to get acquainted with the European legislation for toxic substances denoted REACH, for instance, one can make use of the Internet. A brief presentation of REACH is, however, included in Section 13.7.

1.2 Toolboxes Available Today to Develop an Ecological–Environmental Diagnosis

Massive use of ecological models as an environmental management tool was initiated in the early 1970s. The idea was to answer the question: What is the relationship between the reduction of the impacts on ecosystems and the observable, ecological improvements? (see Figure 1.1). The answer could be used to select the pollution reduction that the society would require and could afford economically. Ecological models were developed in the 1920s by Streeter-Phelps and Lotka Volterra (see, for instance, Jørgensen and Fath, 2011), but a much more consequent use of ecological models started in the 1970s, and many more models of different ecosystems and different pollution problems were developed. In the early 1970s, we had only environmental technology to be used for impact reduction, while today we have more

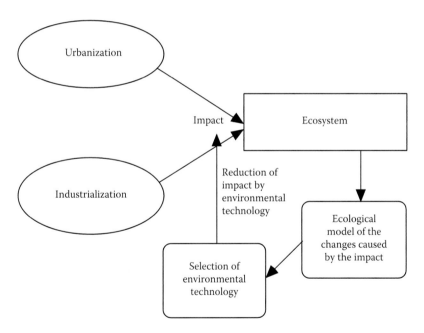

FIGURE 1.1
Illustration of ecological models started in the early 1970s. The models express the changes the impacts (in model language called forcing functions) can cause on the ecosystems and contain a mathematical relationship between forcing functions and the state of the ecosystems described by the use of state variables. The model is used to decide on the required reduction of the forcing functions from a selection of suitable environmental management solutions. In the early 1970s, the only environmental technology available was a toolbox to reduce the impacts on the ecosystems. Today, as revealed in the next section, we have more toolboxes that can be applied to reduce the forcing functions (impacts).

toolboxes as will be presented briefly in the next section. In the late 1970s and early 1980s, several ecotoxicological models were developed (see Jørgensen, 1990) and a procedure to develop environmental risk assessment (ERA) for toxic substances was developed (see Chapter 8).

Today, we have at least a few models available for practically all combinations of ecosystems and environmental problems, including pollution by many toxic substances. The journal *Ecological Modelling* was launched in 1975 with an annual publication of 320 pages and about 20 papers. Today, the journal publishes 20 times as many papers, indicating that ecological modeling has been adopted as a very powerful tool in ecological–environmental management to cover particularly point 4 in the proposed integrated ecological and environmental management procedure.

Ecological models are powerful management tools, but they are not easily developed. They require in most cases good data, which are both resource- and time-consuming. For toxic substances, the model development requires knowledge of the properties of the toxic substances: What is the equilibrium between soil and soil water? How much does the toxic substance biomagnify through the food chain? How quickly will the toxic substance biodegrade to mention a few important properties? The industrialized countries use about 100,000 toxic substances but only the properties for a limited number of toxic substances are known. Therefore, a system denoted QSAR (quantitative structure activity relationships) has been developed. In most cases, the system is able to estimate from the chemical structure with an acceptable accuracy the properties of chemical compounds. Chapter 4 briefly presents the characteristics of QSAR.

If an economically important and environmentally crucial project is the focus, it is often very beneficial to develop a good ecological–environmental model, but if the project is minor, it may be sufficient to consider other solutions. About 25 years ago, it was therefore proposed to use another toolbox that requires less resources to provide a diagnosis, namely, ecological indicators (see, for instance, Costanza et al., 1992). Ecological indicators can be classified as shown in Table 1.1 according to the spectrum from a more detailed or reductionistic view to a system or holistic view (see Jørgensen, 2002). The reductionistic indicators can, for instance, be a chemical compound that causes pollution, including toxic substances (polychlorinated biphenyls [PCB] is mentioned in this context in Table 1.1) or specific species that by their presence or absence can be translated to ecosystem quality. A holistic indicator could, for instance, be a thermodynamic variable or the biodiversity determined for the focal ecosystem. The indicators can either be measured or be determined by the use of a model. In the latter case, the time consumption is not reduced by the use of indicators instead of models, but the models get a more clear focus on one or more specific state variable, namely, the selected indicator, that best describes the problems. In addition, indicators are usually associated with very clear and specific health problems of the ecosystems, which is beneficial in environmental management. Note that the discharge of toxic substances into ecosystems may cause a significant change

TABLE 1.1

Classification of Ecological Indicators

Level	Example
Reductionistic (single) indicators	PCB, species present/absent
Semiholistic indicators	Odum's attributes
Holistic indicators	Biodiversity/ecological network
"Super-holistic" indicators	Thermodynamic indicators as emergy, eco-exergy (work energy) and energy

Source: Adapted from Marques, J.C. et al. 2009. *Marine Pollution Bulletin* 58: 1773–1779; Jørgensen, S.E. and S.N. Nielsen. 2012. *Ecological Indicators* 20: 104–109.

in all ecological health indicators. Chapter 9 is devoted to the application of ecological models as a diagnostic tool and Chapter 10 covers the application of ecological indicators and the assessment of ecosystem services with a focus on toxic substance pollution in the environment.

In the last 15–20 years, the services offered by ecosystems to society have been discussed and an attempt was made to calculate the economic values of these services (Costanza et al., 1997). A diagnosis that would focus on the services that were actually reduced or eliminated due to environmental problems could easily be developed. Another possibility of using ecological services to assess environmental problems and their consequences could be to determine the economic values of the overall ecological services offered by ecosystems and then compare them with what is normal for the type of ecosystems considered. Jørgensen (2010) determined the values of all the services offered by various ecosystems through the use of the ecological holistic indicator eco-exergy (it is the work energy *including* the work energy of the information embodied in the organisms of the ecosystems; for further details, see Jørgensen, 2006). Assessment of the value of ecosystem services can be considered an ecological indicator and the concept of ecosystem services is included in Chapter 10.

An assessment of the ecosystem services is frequently carried out using specific ecological indicators. The indicators are followed by the use of models that can assess the reduced or lost ecological services of ecosystems. The three diagnostic toolboxes are, in other words, closely related and obviously the use of all three toolboxes will give the most complete diagnosis. However, the resources available for environmental management is always limited, which means that it is hardly possible to apply all three toolboxes in all cases, but it is necessary in many cases to make a choice. If an ecological model is developed, anyhow, to be able to give more reliable prognoses, it is natural to apply the developed model and it may be beneficial in addition to the selected one or a few indicators to focus more specifically on a well-defined problem. If a model is not available but a monitoring program has to be developed, it would naturally direct the observations to encompass the state variables that can be applied to assess the indicators that are closely related to the defined health problems. If the society is dependent on specific ecological services of the ecosystem, it would be natural to assess to what extent these services are maintained, reduced, or lost, maybe supplemented with health indicators that are particularly important for the maintenance of these services. The choice of diagnostic toolboxes is therefore a question regarding the available resources and the specific case and problem.

1.3 Toolboxes Available Today to Solve Pollution Problems

The toolbox environmental technology was the only methodological discipline available to solve environmental problems 45–50 years ago when the first green waves started in the mid-1960s. This toolbox was only able to solve the problems of point sources sometimes, but not always, at a very high cost. Today, fortunately, we have additional toolboxes that can solve the problems of the diffuse pollution or find alternative solutions at lower costs when the environmental technology would be too expensive to apply. As for the diagnostic toolboxes, these toolboxes are developed on the basis of new ecological subdisciplines.

To solve environmental problems, we have today four toolboxes:

1. Environmental technology—on a more advanced level than 45–50 years ago
2. Ecological engineering, also denoted as ecotechnology
3. Cleaner production—under this heading, we would also include industrial ecology
4. Environmental legislation

Environmental technology was available through the emergence of the first green waves about 45–50 years ago. An overview of the available methods of environmental technology to solve the pollution problems of toxic substances is presented in Chapter 11. Several new environmental–technological methods have been developed and all the methods have been streamlined and are generally less expensive to apply. However, there is and has been an urgent need for other alternative methods in order to solve the entire spectrum of environmental problems at an acceptable cost. The environmental management today is more complicated than it was 50 years ago because of the many more toolboxes that should be applied to find the optimal solution and because global and regional environmental problems have emerged. The use of toolboxes and the more complex situation today are illustrated in Figure 1.2.

The toolbox containing ecological engineering methods has been developed since the late 1970s. Ecological engineering is defined as the design of sustainable ecosystems that integrate human society with its natural environment

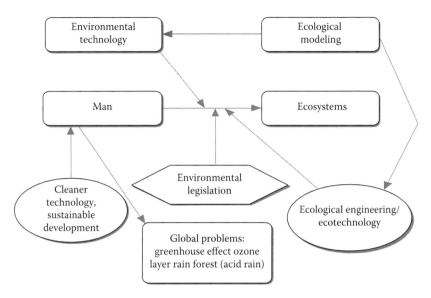

FIGURE 1.2
Conceptual diagram of the present-day complex ecological–environmental management, where there are various toolboxes available to solve the problems and where the problems are local, regional, and global. (Adapted from Jørgensen, S.E. 2013. *Encyclopedia of Environmental Management.* CRC Press, Boca Raton, Four volumes, 3280pp.)

for the benefit of both (Mitsch and Jørgensen, 2004). It is an engineering discipline that operates in ecosystems, which implies that it is based on both design principles and ecology. The toolbox contains four classes of tools:

1. Tools that are based on the use of natural ecosystems to solve environmental problems (for instance, the use of wetland to treat agricultural drainage water)
2. Tools that are based on imitation of natural ecosystems (for instance, construction of wetlands to treat wastewater)
3. Tools that are applied to restore ecosystems (for instance, restoration of lakes by the use of biomanipulation)
4. Ecological planning of the landscape (for instance, the use of agroforestry)

The introduction of ecological engineering has made it possible to solve many problems that environmental technology could not solve, first of all nonpoint pollution problems and a fast restoration of deteriorated ecosystems. Further details on ecological engineering are presented in Chapter 12 with emphasis on the use of ecological engineering to solve toxic substance problems.

As the environmental legislation has been tightened, it has become increasingly more expensive to treat industrial emissions, and the industry has considered whether it is possible to reduce emission by other methods at a lower cost. That has led to the development of what is called cleaner production, which means to produce the same product with a new method that would give reduced emission and therefore cost less for pollution treatment. New production methods have been developed through the use of innovative technology that has created a completely new way to produce the same product with less environmental problems. In case of pollution problems associated with the use of toxic substances in the production, it is obvious that a new production method, not based on the use of toxic substances characterizing the "old" production method, eliminated the environmental problem, often at relatively low costs. In this context, it is also possible to consider production methods that are based on replacing a toxic substance with a less, but still, toxic substance. Other emission reductions have been developed through the use of ecological principles on industrial processes, for instance, recycling and reuse, which can often be used for toxic substances. In many cases, it has also been possible to achieve a reduction of the environmental problems by identification of unnecessary waste, including toxic substance waste, which otherwise would need an environmental–technological treatment.

Some environmental problems, however, cannot be solved without stricter environmental legislation, and for some problems, a global agreement may be needed to achieve a proper solution; for instance, as in the case of phasing out the use of freon to stop or reduce the destruction of the ozone layer. Note that environmental legislation also requires an ecological insight to assess the required emission reduction needed through the introduction of environmental legislation. As mentioned, environmental legislation is only covered briefly in this book, but reference is given to Jørgensen et al. (2015) and to the details of the European legislation on REACH, which is close to the legislation in other industrialized countries, including the United States, Canada, Japan, Australia, and New Zealand.

Today, through the four toolboxes with environmental management solutions, we have a possibility to solve any environmental problem, often at a moderate cost and sometimes even at a cost that makes it profitable to solve the problem properly, because of reduced use of resources. As is the case for diagnostic toolboxes, the toolboxes with problem solution tools are also rooted in recently developed ecological subdisciplines that are named after the tools: ecological engineering, cleaner technology, and environmental legislation.

1.4 Follow the Recovery Process

Environmental management is only complete if the environmental problem and the ecosystem are followed carefully after the toolboxes have been applied. It is usually not a problem because it is a question of providing the observations needed to follow the prognoses of the following:

a. Eventually developed ecological model
b. Selected ecological indicators
c. Recovery of the ecological services of the ecosystem (which can be done by focusing on a specific service or on the total values of all the ecological services offered by the ecosystem)

1.5 Conclusions about Integrated Environmental Management

From the review of the up-to-date integrated environmental management, it is possible to conclude the following:

1. Follow all the seven recommended steps, using a holistic approach. Integrate the knowledge about the problem(s), the ecosystems involved, and the sources with a good diagnosis and a combination of solution methods to reach an optimum environmental management. A good knowledge of systems ecology is important in this context.

2. Consider using all the three diagnostic toolboxes, but use at least the diagnostic toolbox that fits into the problem, ecosystem, and the available observations. A good diagnosis is an indispensable step in integrated management.

3. The three diagnostic toolboxes can also be applied to follow the development of an environmental–ecological problem, including an eventually recovery process.

4. Eco-exergy—work energy including the work energy of information—is a useful indicator as it expresses sustainability and thereby the total amount of ecological services offered by an ecosystem.

5. Integrated environmental management based on the three diagnostic toolboxes and considering all sources of the problem may require the use of all four "problem-solving" toolboxes:
   - Environmental technology
   - Ecotechnology
   - Cleaner technology, including industrial ecology
   - Environmental legislation

A combination (integrated use of) of the four toolboxes will often be able to offer the best and most cost-moderate solution.

Integrated up-to-date environmental management requires the use of the seven presented toolboxes and would not be possible if these toolboxes were not developed as a result of recently emergent ecological subdisciplines: ecological modeling, ecological engineering, application of ecological indicators, cleaner technology, and industrial ecology. These ecological subdisciplines are therefore crucial for the present-day environmental management and they form an indispensable bridge between ecology and environmental management—between the basic science of ecology and its application in practical environmental management. This conceptual bridge symbolizing the close and integrated cooperation between the ecological subdisciplines and environmental management is illustrated in Figure 1.3. The outlines of the book follow to a certain extent the seven steps of an integrated, holistic environmental management. The next chapter focuses on the determination of the sources of toxic substances while Chapter 3 deals with the properties of toxic substances, which are crucial for the effects that are observed in ecosystems. As we unfortunately only know the properties of a limited number of the 100,000 toxic substances that are applied in industrialized countries, it is necessary in many cases to use estimation methods (QSAR; see above) as an acceptable solution. Chapter 4 gives a brief overview of the available estimation methods. Chapter 5 looks

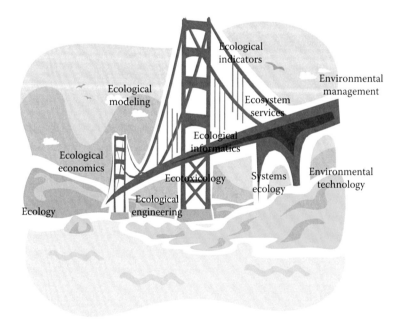

FIGURE 1.3
(**See color insert.**) A conceptual bridge illustrating the close and integrated cooperation between subdisciplines of ecology and environmental management, which is a prerequisite for an up-to-date and holistic solution of the environmental problem.

into the cycles of a number of elements, because it provides an understanding of the global distribution and processes of toxic substances. Chapter 6 gives an overview of the toxic substances that threaten environmental quality. Chemical compounds, including toxic substances, participate in physical–chemical processes in the environment and it is very important to be able to determine to what extent the chemical compounds react in the environment, as the impact on the environment is dependent on which chemical compounds are present in the ecosystems in what concentration. It should be emphasized that the toxicity of toxic compounds are highly dependent on these reactions. Chapter 7 is devoted to the physical–chemical calculations of reactions in the environment. Chapters 8 through 10, as already mentioned, cover diagnostic toolboxes, ERA, ecological modeling, and ecological indicators. Chapters 11 through 13 present the toolboxes that can be used to find solutions to environmental problems due to discharge of toxic substances into ecosystems. Chapter 14 deals with how to develop and apply integrated environmental management to solve toxic substance problems. The chapter tries to answer the following crucial questions:

1. Why is it necessary to apply an integrated approach?
2. How is an integrated environmental management approach applied?
3. What are the illustrative examples of the solution of toxic substance pollution?

The last chapter, Chapter 15, presents the summaries and conclusions.

References

Costanza, R., B.G. Norton, and B.D. Haskell. 1992. *Ecosystem Health: New Goals for Environmental Management.* Island Press, Washington, 270pp.

Costanza, R. et al. 1997. The value of the world's ecosystem services and natural capital. *Nature* 387: 252–260.

Jørgensen, S.E. 1990. *Modelling in Ecotoxicology.* Elsevier, Amsterdam, 156pp.

Jørgensen, S.E. 2002. *Integration of Ecosystem Theories: A Pattern.* Kluwer, Dordrecht, 386pp.

Jørgensen, S.E. 2006. *Eco-Exergy as Sustainability.* WIT, Southampton, 220pp.

Jørgensen, S.E. 2010. Ecosystem services, sustainability and thermodynamic indicators. *Ecological Complexity* 7: 311–313.

Jørgensen, S.E. 2013. *Encyclopedia of Environmental Management.* CRC Press, Boca Raton, Four volumes, 3280pp.

Jørgensen, S.E. and B. Fath. 2011. *Fundamentals of Ecological Modelling.* 4th edition. Elsevier, Amsterdam, 400pp.

Jørgensen, S.E., J.C. Marques, and S.N. Nielsen. 2015. *Integrated Environmental Management: A Transdisciplinary Approach.* CRC Press, Boca Raton, 380pp.

Jørgensen, S.E. and S.N. Nielsen. 2012. Tool boxes for an integrated ecological and environmental management. *Ecological Indicators* 20: 104–109.

Marques, J.C. et al. 2009. Ecological sustainability trigon. *Marine Pollution Bulletin* 58: 1773–1779.

Mitsch, W.J. and S.E. Jørgensen. 2004. *Ecological Engineering and Ecosystem Restoration.* John Wiley, New York, 410pp.

2

Sources of Toxic Substance Pollution

2.1 Sources of Pollution

In this chapter, we distinguish between point sources and nonpoint sources and identify 13 forms of pollution. The impact of pollution on the environment is mainly due to three factors: increasing population, increasing human consumption and production, and use of technology, which increases or decreases the impact depending on the type of applied technology.

Pollution is defined as the introduction of contaminants into the natural environment, which causes adverse and undesirable changes in the ecosystem. Pollution can either be matter, such as chemical substances, or energy, such as noise, heat, or light. Pollutants can be either human-controlled—meaning as a result of human activities—or naturally occurring—resulting from such natural events as a volcanic eruption. Both human-controlled and naturally occurring pollution contain toxic substances. It is very important to distinguish between point source and nonpoint source pollution because the two sources will in most cases require different toolboxes to be eliminated. Environmental technology is developed to solve point source pollution and will in most cases not be applicable to nonpoint pollution, while ecological engineering is better able to cope with nonpoint pollution. Therefore, it is important to consider the entire spectrum of possible solutions for each of the sources. Point sources always have a minor or major concentration of toxic substances and include, for instance, discharge from wastewater treatment plants, discharge of wastewater from industries, smoke from a chimney, or ash from a solid waste incineration plan. Nonpoint sources are diffuse and their examples include agricultural pollution caused by drainage water containing pesticides and nutrients or by storm water. Typical analyses of the two important nonpoint pollution sources, storm water and agricultural drainage water, are shown in Table 2.1. They contain toxic substances but it is hardly possible to present a general concentration of toxic substances in storm water and agricultural drainage water, as the spectrum of toxic contaminants is very wide. Storm water contains toxic substances originated from air pollutants and washout by precipitation, from soil and from solid waste, debris, and waste.

TABLE 2.1

Pollution from Urban and Agricultural Runoff

Constituent	Storm Water	Agricultural Drainage Water
Suspended solids (mg/L)	5–1200	5–5000
Chemical oxygen demand (COD)[a] (mg/L)	20–700	5–1000
Biological oxygen demand (BOD5)[b] (mg/L)	1–200	1–300
Total phosphorus (mg/L)	0.02–8	0.1–2.0
Total nitrogen (mg/L)	0.3–8	0.5–10
Chlorides (mg/L)	1–35	–

[a] COD is oxygen demand determined by use of chromate.
[b] BOD5 is oxygen used by microbiological decomposition over a period of 5 days.

Agricultural drainage water obviously contains toxic substances used in agriculture, that is, pesticides, veterinary medicine, and toxic substances contained in fertilizers.

Pollution is often classified by the sphere that is polluted. When the hydrosphere is contaminated, it is denoted as water pollution. When the atmosphere is contaminated, it is called air pollution. Terrestrial pollution or soil pollution is the pollution of the lithosphere. There are, however, often pollutants that are contaminants in all the three spheres, and in the biosphere, where the effect often is most radical and most important.

The major forms of pollution are covered in the following list (with an indication where toxic substances are relevant for the pollution problem) (Jørgensen and Nielsen, 2012; Jørgensen et al., 2015):

1. Common gaseous pollutants include carbon dioxide, carbon monoxide, sulfur dioxide, chlorofluorocarbons (CFCs), and nitrogen oxides produced by industry, power plants, and motor vehicles. They are all toxic except carbon dioxide, which is, however, the most important greenhouse gas.

2. Photochemical ozone and smog are created when nitrogen oxides and hydrocarbons react to sunlight. The source is industry and vehicles. All the components are toxic.

3. Particulate matter, or fine dust, is characterized by size, contaminating the atmosphere and deteriorating air quality. Particulate matter contains toxic substances, including the toxic gaseous contaminants mentioned above under point 1, as they are adsorbed by particulate matter.

4. Light pollution includes light trespass, over-illumination, use of laser, and astronomical interference. No toxic matter is involved.

5. Noise pollution includes roadway noise, aircraft noise, industrial noise, as well as high-intensity sonar. No toxic matter is involved.

6. Radioactive contamination results from nuclear power generation and nuclear weapons research, manufacture, and deployment and from the general use of radioactive isotopes. Radioactive contaminants are considered toxic substances.

7. Thermal pollution is a temperature change in natural water bodies caused mainly by the use of water as coolant in a power plant. No toxic matter is involved.

8. Soil contamination occurs when chemicals are released by spill or underground leakage. The possible contaminants are listed below as toxic substance pollution.

9. Visual pollution, which can refer to the presence of overhead power lines, motorway billboards, scarred landforms (as from strip mining), open storage of trash, littering, municipal solid waste, or space debris, could also be named landscape pollution. No toxic matter is involved.

10. Pollution of water by organic matter causes oxygen depletion due to the use of oxygen for mainly biological decomposition of organic matter. Note that this pollution is a result of two sources: nonpoint sources (agricultural drainage water and storm water) and insufficiently treated wastewater. Toxic substances decompose, too, although often slower than nontoxic organic components, meaning that toxic substances contribute to oxygen depletion anyhow by aerobic microbiological decomposition processes.

11. Pollution of aquatic ecosystems (surface waters) by nutrients causing eutrophication. The nutrients cause extensive growth of algae (or other plants), which reduces the transparency of water and causes oxygen depletion in the bottom water in the fall and winter. Note that this pollution is a result of mainly two sources: nonpoint sources (agricultural drainage water and storm water) and insufficiently treated wastewater. Toxic substances are not involved directly, but some blue-green algae may excrete toxic substances that may lead to prohibition of bathing and swimming in lakes.

12. Groundwater pollution covering the disposal and leaching of nutrients and toxic substances to groundwater. Note that this pollution is often the result of an uncontrolled use of nutrients and pesticides in agricultural production. Toxic substances in groundwater threaten human health, as groundwater is widely used for the production of drinking water.

13. Toxic substance pollution, which can cause adverse effects in all spheres, including the lithosphere (see point 8) and the biosphere. Later in this book, an overview and classification of toxic substances will be presented.

In accordance with the proposed seven-step procedure in Chapter 1, it is important to quantify *all* possible pollution sources of toxic substances. Quantification is the key because only with the knowledge of the quantities of all the sources can we, with the diagnostic tools (see Chapters 8 through 10), find which sources are necessary to be eliminated or reduced to be able to obtain the needed and desired impact reduction. Mass conservation principles are often used to determine the quantities of the various pollution sources of toxic substances. From emission at the discharge point to impact on the ecosystem, some processes may have changed the concentrations and amounts of the polluting components. For instance, from the emission of gases from a chimney to the dry deposition in an ecosystem, chemical transformations and dilution processes may have occurred. Chapter 7 will focus on these processes, which are necessary to consider to be able to determine the effect of an emission.

All sources of pollution and problems are rooted in urbanization, industrialization, inappropriate environmental planning or application of available solutions, and natural disasters. It is, however, due to human activities that most pollution problems occur. The increased impact of environmental problems is dominantly due to increased urbanization and industrialization. In the last few decades, we have observed a distinct increase in pollution. Many examples illustrate these observations. The concentration of many toxic substances has increased in soil, water, and biosphere. The concentration of carbon dioxide, sulfur dioxide, and other gaseous pollutants has increased drastically in the atmosphere. Oxygen depletion has been recorded in many rivers, and many aquatic ecosystems are suffering from eutrophication with high concentrations of nitrogen and phosphorus. What has caused this rapid increase in pollution? The answer is not simple, but Ehrlich and Raven (1969) have proposed a crude but useful model. They claim that the total environmental impact, I, is mainly determined by three multiplicative factors: the population size, P, the consumption per capita, C, and the environmental impact per unit of consumption, which is dependent on the technology, T:

$$I = P * C * T \tag{2.1}$$

It is understandable that the impact is proportional to P, as twice as many people with the same consumption and technology of course will cause twice as large an impact. Similarly, if we consume and produce twice as much per capita, the impact will inevitably be twice as large. The factor T is more difficult to find and manage, because on the one side, for instance, a technological development leading to the production of more harmful chemicals increases the impact, while the development of more recycling and reuse will lead to less impact. Development of, for instance, renewable energy will decrease carbon dioxide emission and thereby decrease the total impact. P has an increase a factor of 2 from 1960 to 2000—from 3 to 6 billion.

The global gross national product (GNP) per capita, which reflects consumption and production, has also increased by a factor about 2 from 1960 to 2000. This means that the global impact has increased four times from 1960 to 2000, presuming that T has not changed. We are producing more harmful chemicals today than 50 years ago, but we have also developed technology that emits less pollution and enables more recycling. However, an increased impact of four times (400%) in 40 years can only be considered as a very rapid increase, which calls for pollution abatement. T is highly influenced by the use of toxic substances. If the controlled use of toxic substances is increased, T will decrease. If we replace toxic substances by less toxic substances, which is the idea in the application of cleaner technology (see Chapter 12), T will decrease. If the emission of toxic gases from vehicles is reduced through the use of catalysts, T decreases, but if we use more pesticides and veterinary medicine in agriculture, T increases. Several changes in industrialized societies in the last 40 years have increased T, while several others have decreased T. As the population and GNP per capita increase, it is however necessary to make a significant effort to decrease T to at least try to compensate partially for the increased I. Through a much wider application of recirculation, cleaner technology, and controlled use of chemicals in general, it is indeed possible to reduce T significantly, which is absolutely needed.

There are 24 elements that are needed for the entire biosphere. All 24 elements are not necessarily essential for *all* organisms, but each organism will require most of these 24 elements for their growth. These elements are threshold agents (see Jørgensen, 2000). They are needed in a certain concentration for the growth of the organisms, but if they are present in too high a concentration, they may be harmful for the environment. A typical example is the eutrophication problem, which is the result of an excessive concentration of nutrients, but the nutrients (with emphasis on nitrogen, carbon, phosphorus, and sulfur) are absolutely necessary for all living organisms. Nonthreshold agents or gradual agents are harmful for the environment in practically any concentration. The harmful effect may be proportional to the concentration or there may be another relationship between the effect and the concentration, for instance, the harmful effect is proportional to the concentration in exponent 2.

Figure 2.1 illustrates the harmful effect versus the concentration for threshold and nonthreshold agents.

Heavy metals are an important contributor to the problem of soil pollution, and a more detailed overview of this will be presented in Chapter 6. Three heavy metals require particular environmental attention: mercury, lead, and cadmium, because they cause very serious environmental problems.

It is recommended to start the management by providing the needed survey of the sources by following these points:

1. List all the sources—include at this stage also all suspected sources.
2. Indicate whether the sources are point sources or nonpoint sources.

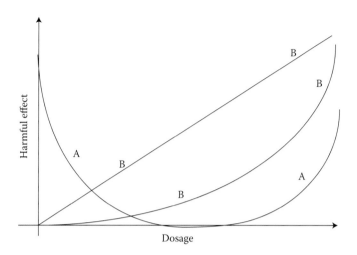

FIGURE 2.1
The harmful effect is plotted versus the dosage or concentration for (A) threshold agents and (B) nonthreshold agents.

3. Indicate whether the sources are pulsing or continuous, or even better give the sources as f (time).

4. Quantify all the sources.

5. What is the impact on the ecosystem (nature) or human health as a function of the magnitude of the environmental problem? Is the impact proportional to the magnitude of the environmental problem or how can we describe the impact as f (magnitude of the problem)?

6. The impact is often a result, determined by an interaction of various compounds coming from different sources. The effects of toxic substances can, as an example, reinforce each other significantly. The effect is not at all just additive.

7. The duration of the source and thereby the corresponding impact is often decisive for the final effect of the environmental problem. It is the core topic of the next section.

2.2 Distribution in Time and Space

The harmful effect of a pollution source is not only dependent on the amount and type of pollutants but also on the distribution of the pollution in time and space. Different parts of the environment have different sensitivity to specific pollutants and the duration of the pollution is furthermore decisive for the effect. This section will focus on the importance of the pollution distribution in time and space. Calculations—for instance, by fugacity models—can be accomplished. Space and time are, however, interrelated. How is large area or volume affected in a long time? It is an obvious question that we need to answer to fully understand the harmful effect of pollution. The various adverse effects and the ecological hierarchy have different scales in time and space, which must be considered in a proper environmental analysis. Or expressed differently, the larger the affected area or volume, the longer will be the time for recovery by nature after a pollution case. The interrelationship between area and recovery time is shown in Figure 2.2 (Jørgensen, 2000, 2012). It implies that pollution problems increase more than proportional to the area or volume because both the area or volume affected and the recovery time needed increase.

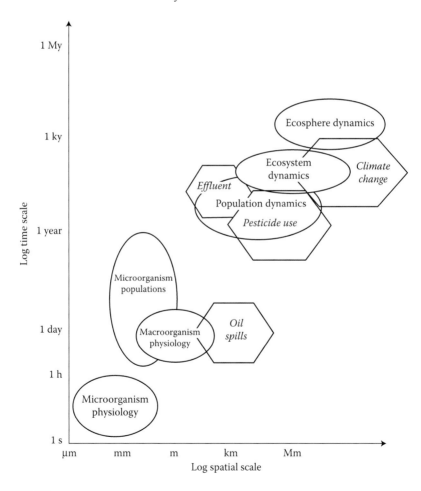

FIGURE 2.2
The spatial and time scale for various hazards (hexagons, italic) and for the various levels of the ecological hierarchy (circles, nonitalic).

To obtain the complete image of a pollution case, the chemical, physical, and environmental processes of the focal compounds should also be taken into account. A number of processes are possible for the pollutants emitted to the environment, and the processes are also dependent on the distribution in time and space. The most important processes are dilution; transport by air, water, or soil; biodegradation; hydrolysis; photolysis; dissolution; evaporation; settling; precipitation; oxidation; reduction; and formation of complexes. It is, however, very important to consider all the case-relevant processes because the compounds formed may have a completely different effect and environmental hazard than the original emitted compounds. It is therefore crucial to determine the concentrations of the formed compounds. A good overview is therefore required in many cases due to the complexity of the problem and the development of an ecotoxicological/environmental and chemical model. Chapters 7 through 10 will deal with the quantification of chemical and physical processes and the development of environmental impact assessments (EIAs), models, and ecological indicators to obtain the needed overview of the pollution problem. These four chapters cannot cover the physical–chemical calculations and the development of ecotoxicological models in detail, but a more detailed coverage of ecotoxicological models can be found in Jørgensen and Fath (2011). In this section, the development of fugacity models is briefly presented because the most simple fugacity model gives the result of the first rough spatial distribution based upon the sources. Three higher levels of fugacity models are however able to consider some of the environmental processes, particularly the physical transport processes.

Several important properties of the chemical compounds determine both the effect—meaning a translation of the resulting concentration to hazard—and the processes that the chemicals undergo in the environment, which must be calculated based upon the properties either by direct calculations or by development of models. An overview of the properties is presented in the next chapter.

Another important question is which part of the environment is mostly affected, considering the different sensitivities to different pollutants. The answer to this question will require knowledge of the above-mentioned spatial distribution of the pollutants. An overview of the spatial distribution can be obtained by the use of a so-called fugacity model that has just been mentioned above. This model type, originally developed mainly by Mackay (1991), has wide application in environmental chemistry with many different models, developed by different authors (see SETAC, 1995).

These models are based on the concept of fugacity, $f = c/Z$, where c is the concentration in the considered phase and Z is the fugacity capacity (measured in moles/m^3 Pa or moles/L atm). Fugacity is defined as the escaping tendency, and has the units of pressure (atmosphere or Pa) identical to the partial pressure of ideal gases. By equilibrium between two phases, the fugacity of the two phases is equal. If the two Zs are known, then it is possible to calculate the concentrations in the two phases. If there is no equilibrium, then the rate of transfer from one phase to the other is proportional to the difference in fugacity.

If the equation for ideal gases can be applied, we have $pV = nRT$, where n is the number of moles, R is the gas constant $= 8.314$ Pa m^3/mole K, and T is the absolute temperature. This leads to $p = c * R * T$, and

$$c = \frac{p}{RT} = \frac{f}{(RT)} \tag{2.2}$$

By acceptable approximation (application of the equation for ideal gases and the activity is equal to the concentration), the fugacity capacity in air is

$$Z_a = \frac{1}{RT} \tag{2.3}$$

At equilibrium between water and air, the fugacity is the same in the two phases, as already mentioned:

$$c_a Z_a = c_w Z_w \tag{2.4}$$

where w is used as index for water.

Based on Henry's law, $p = He*y$, where, as used above, $p = c_a RT$ and $y = c_w/(c_w + [H_2O])$, we can find the distribution between air and water. The concentration of water in water is with good approximation $1000/18 \gg c_w$, which means that we get $p = c_a RT = He\ y = He\ c_w/(c_w + [H_2O]) = He\ c_w\ 18/1000$. Equation 2.4 yields

$$\frac{c_a}{c_w} = \frac{Z_a}{Z_w} = \frac{18\ H}{1000\ RT} \tag{2.5}$$

It implies that $Z_w = 1000/18He$.

Similarly, the distribution between water and soil (index s) can be applied to find the fugacity capacity of soil:

$$\frac{c_s}{c_w} = \frac{Z_s}{Z_w} = K_{ac} \tag{2.6}$$

Z_s is calculated as $Z_w * K_{ac} = 1000\ K_{ac}/18He$. In a parallel manner, Z_o, the fugacity capacity for octanol, can be obtained as $1000\ He\ K_{ow}/18$ and the fugacity capacity for biota, Z_b, as $1000\ He\ BCF/18$. Table 2.2 gives an overview of the obtained fugacity capacities in mole/L atm. $R = 0.0820$ atm L/(mole K), when these units are applied. If m^3 is the unit for volume and Pa is the unit for pressure, then we get 1 atm $= 101,325$ Pa and 1 L $= 1/1000$ m^3. It implies that

TABLE 2.2

Fugacity Capacity

Phase	In moles/L atm
Atmosphere	$1/RT$ ($R = 0.0820$)
Hydrosphere	$1000/He\ 18$
Lithosphere (soil)	$1000\ K_{oc}/18\ He$
Octanol	$1000\ K_{ow}/18\ He$
Biota	$1000\ BCF/18\ He$

If the unit moles/m³ Pa is required, divide by 101.325.

R has the units J/mole K corresponding to the value $0.082 \times 101,325/1000 =$ 8.3 J/(mole K). Figure 2.3 shows a conceptual diagram of the most simple fugacity model and Table 2.2 summarizes the expression used.

Multimedia models or fugacity models can be applied on four levels. An equilibrium distribution (level 1) is found from the known fugacity capacities and equal fugacities in all spheres. If advection and chemical reactions must be included in one or more phases, but the equilibrium is still valid, then we have level 2. The fugacities are still the same in all phases. Level 3 presumes steady state but no equilibrium between the phases.

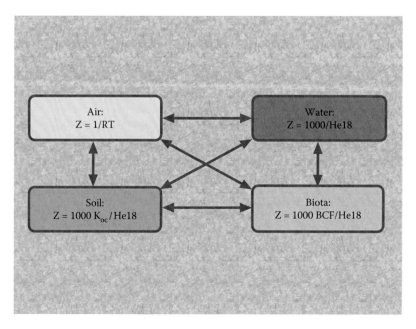

FIGURE 2.3
(**See color insert.**) A fugacity model gives an overview of the distribution of chemical (toxic) compounds among the spheres: atmosphere, hydrosphere, lithosphere, and biosphere. The applied fugacity expressions are shown in the figure.

Transfer between the phases is therefore taking place. The transfer rate is proportional to the fugacity difference between the two phases. Level 4 is a dynamic version of level 3, which implies that all concentrations and possibly also the emissions change over time.

If the total emission in all phases is denoted M, then we have

$$M = \sum c_i V_i = f \sum Z_i V_i \qquad (2.7)$$

where c_i, V_i, and Z_i are concentration, volume, and fugacity capacity of sphere number i. Levels 1 and 2 are usually sufficient to calculate the environmental risk of a chemical. For level 1 calculations, the fugacity capacities obtained from Table 2.2 are applied to find f because the total emission and the volumes of the spheres are known. The concentrations are then easily determined from $c_i = fZ_i$. The amounts in the spheres are found from the concentrations times the volumes of the spheres. Example 2.1 presents these calculations.

EXAMPLE 2.1

A chemical compound has a molecular weight of 200 g/mole and a water solubility of 20 mg/L, which gives a vapor pressure of 1 Pa. The distribution coefficient of octanol–water is 10,000 and $K_{ac} = 4000$. How will an emission of 1000 moles be distributed in a region with an atmosphere of 6×10^8 m³, a hydrosphere of 6×10^6 m³, a lithosphere of 50.000 m³ with a specific gravity of 1.5 kg/L, and an organic carbon content of 10%. Biota (fish) is estimated to be 10 m³ (specific gravity 1.00 kg/L and a lipid content of 5%). The temperature is presumed to be 20°C.

Solution

Fugacity capacities:

$Z_a = 1/RT = 1/8.314 * 293 = 0.00041$ moles/m³ Pa
$Z_w = (20/200)/1 = 0.1$ moles/m³ Pa
$Z_s = 0.1 \times 0.1 \times 4000 = 40$ moles/m³ Pa
$Z_{biota} = 0.1 \times 0.05 \times 10,000 = 50$ moles/m³ Pa
$\_ Z_i V_i = 0.00041 \times 6 \times 10^8 + 0.1 \times 6 \times 10^6 + 40 \times 50.000 + 10 \times 50 = 2,846,500$ moles/Pa
$f = M/\Sigma Z_i V_i = 1000/2,846,500 = 3.51 \times 10^{-4}$

Concentrations:

$c_a = fZ_a = 3.51 \times 10^{-4} \times 0.00041 = 1.44 \times 10^{-7}$ moles/m³
$c_w = fZ_w = 3.51 \times 10^{-4} \times 0.1 = 3.51 \times 10^{-5}$ moles/m³
$c_s = fZ_s = 3.51 \times 10^{-4} \times 40 = 1.404 \times 10^{-2}$ moles/m³
$c_{biota} = fZ_{biota} = 3.51 \times 10^{-4} \times 50 = 1.755 \times 10^{-2}$ moles/m³

Amounts:

$$M_a = c_a V_a = 1.44 \times 10^{-7} \text{ moles/m}^3 \times 6 \times 10^8 \text{ m}^3 = 86 \text{ moles}$$
$$M_w = c_w V_w = 3.51 \times 10^{-5} \text{ moles/m}^3 \times 6 \times 10^6 \text{ m}^3 = 211 \text{ moles}$$
$$M_s = c_s V_s = 1.404 \times 10^{-2} \text{ moles/m}^3 \times 50.000 \text{ m}^3 = 702 \text{ moles}$$
$$M_{biota} = c_{biota} V_{biota} = 1.755 \times 10^{-2} \text{ moles/m}^3 \times 10 \text{ m}^3 = 0.2 \text{ moles}$$

The sum of the four amounts is 999.2, which is in good accordance with the total emission of 1000 moles.

Fugacity models, level 2, presumes a steady-state situation, but with a continuous advection to and from the phases and a continuous reaction (decomposition) of the considered chemical. Steady state implies that input = output + decomposition. The following equation is therefore valid:

$$E + \sum Gin_i \times c_{i\,ind} = \sum Gout_i \times c_i + \sum V_i c_i k_i \qquad (2.8)$$

where E is the emission, $Gin_i i$ is the advection into the phase i, $c_{i\,ind}$ is the concentration in the inflow, $Gout_i$ is the outflow by advection, c_i is the concentration in the phase, and $V_i c_i k_i$ is the reaction of the considered component in phase i. As $c_i = f Z_i$, we get the following equation:

$$E + \sum Gin_i c_{i\,ind} = f \left(\sum Gout_i Z_i + \sum V_i c_i k_i \right)$$

where f is the total amount of the component going into phase i divided by $(\sum Gout_i Z_i + \sum V_i Z_i k_i)$. We can often presume that $Gin_i = Gout_i$ denote G_i. The concentration in the phase is as usual $f Z_i$. The amount is correspondingly the concentrations in the phase multiplied by the volume. The turnover rate of the compound in phase i is $f(G_i Z_i + V_i c_i k_i)$.

The concentrations of toxic components in organisms have of course a particular focus because the effect on the biosphere has obviously a particular attention.

A wide variety of terms is used in an inconsistent and confusing manner to describe uptake and retention of pollutants by organisms using different paths and mechanisms. However, three terms are now widely applied and accepted for these processes:

1. *Bioaccumulation* is the uptake and retention of pollutants by organisms via *any* mechanism or pathway. It implies that both direct uptake from air and water and uptake from food are included. The bioaccumulation factor therefore covers the concentration in the organisms divided by the concentration in the air or water and the contamination of the food is included.

2. *Bioconcentration* is uptake and retention of pollutants by organisms directly from water through gills or epithelial tissue. This process is often described by means of a concentration factor, indicating the ratio of the concentration in the organism divided by the concentration in the medium. In this case, the contamination of the food is not included.

3. *Biomagnification* is the process whereby pollutants are passed from one trophic level to another and it exhibits increasing concentrations in organisms related to their trophic status. The concentration in top carnivorous organisms may often be several magnitudes higher than in the medium.

The duration of the pollution source and the distribution of pollutants in time are, as already emphasized, important factors. For toxic substances, the LC_{10}-value, it means the lethal concentration causing 10% mortality among the test animals, is for instance determined for 48 h and 96 h duration of the experiments and the value is of course different for the two different durations of the experiment.

The toxicity of a compound varies with exposure time. Time/effect relationships are of course important in understanding toxic effect. This has already been mentioned above for air pollutants where it is reflected in the air-quality standards. Figure 2.4 shows a toxicity curve of log exposure time versus log LC_{50}. Threshold LC_{50} indicates when the curve becomes asymptotic to the time axis. The *threshold* LC_{50} is usually (should be) magnitudes greater than the concentration found in nature or the threshold values reflected in environmental standards. Figure 2.5 gives another example of

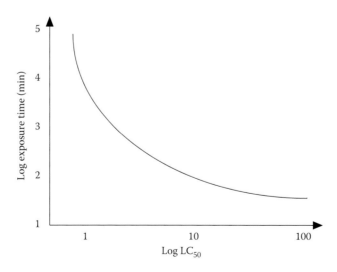

FIGURE 2.4
Log exposure time versus log LC_{50} (mg copper ions/L). The plot indicates LC_{50} for green algae in hard water at pH = 7.0.

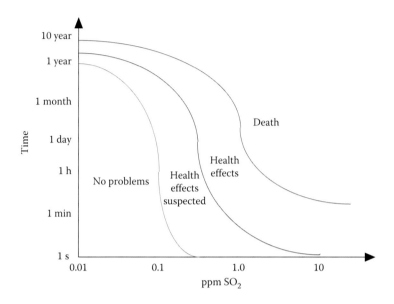

FIGURE 2.5
Effects of sulfur pollution on human health.

the importance of the exposure time (for details, see Jørgensen, 2000). The exposure duration is considered by the standards used for the air quality. The standards are different for annual geometric means, maximum 24-h concentrations, maximum 1-h concentration, and so on.

References

Ehrlich, P.R. and P.H. Raven. 1969. Differentiation of population. *Science* 165: 1228–1232.

Jørgensen, S.E. 2000. *Pollution Abatement*. Elsevier, Amsterdam, 488pp.

Jørgensen, S.E. 2012. *Introduction to Systems Ecology*. CRC Press, Boca Raton, 342pp.

Jørgensen, S.E. and B. Fath. 2011. *Fundamentals of Ecological Modelling*. 4th edition. Elsevier, Amsterdam, 400pp.

Jørgensen, S.E., J.C. Marques, and S.N. Nielsen. 2015. *Integrated Environmental Management: A Transdisciplinary Approach*. CRC Press, Boca Raton, 380pp.

Jørgensen, S.E. and S.N. Nielsen. 2012. Tool boxes for an integrated ecological and environmental management. *Ecological Indicators* 20: 104–109.

Mackay, D. 1991. *Multimedia Environmental Models: The Fugacity Approach*. Lewis Publishers, Boca Raton, 257pp.

SETAC. 1995. *The Multi-Media Fate Model: A Vital Tool for Predicting the Fate of Chemicals*. SETAC PRESS, Pensacola.

3

Properties of Toxic Substances

3.1 Properties Determining Distribution in the Environment

Slightly more than 100,000 chemicals are produced in such an amount that they threaten or may threaten the environment. They cover a wide range of applications: household chemicals, detergents, cosmetics, medicines, dye stuffs, pesticides, intermediate chemicals, auxiliary chemicals in other industries, additives to a wide range of products, chemicals for water treatment, and so on. They are viewed as mostly indispensable to modern society, resulting in an increase of about 40-fold in the production of chemicals during the last four to five decades. A minor or even sometimes a major proportion of these chemicals reaches the environment through their production, transport, application, or disposal. In addition, the production or use of chemicals may cause more or less unforeseen waste or by-products, for instance, chlorocompounds from the use of chlorine for disinfection. The conflict lies in the fact that humans would like to have the benefits of using chemicals but cannot accept their harmful effects, which raises several urgent questions that we have already mentioned in this book. These questions cannot be answered without at least partial application of models, and we cannot develop models without knowing the most important parameters, at least within some ranges. OECD has made a review of the properties that we should know for all chemicals. We need to know the boiling point and melting point to identify the form in which the chemical (as solid, liquid, or gas) is found in the environment. We must know the distribution of chemicals in the five spheres: hydrosphere, atmosphere, lithosphere, biosphere, and technosphere, as mentioned in Chapter 2; their solubility in water; their lipid/water partition coefficient, denoted by K_{ow}; Henry's constant; their vapor pressure; their rate of degradation by hydrolysis, photolysis, chemical oxidation, and microbiological processes; and the adsorption equilibrium between water and soil—all as a function of temperature. We need to discover the interactions between living organisms and chemicals, which implies that we should know the biological concentration factor (BCF), that is, the ratio of the concentration in an organism to the concentration in the media, biomagnification (see Chapter 2), uptake rate and excretion rate by the involved organisms, and where in the organisms the chemicals will be concentrated, not only for one organism but for a wide range of organisms. Table 3.1 gives an overview of the most relevant physical–chemical properties of organic compounds and their interpretation with respect to their behavior in the environment, which should be reflected in their quantification and possibly the developed models. A complete knowledge of all consequences of the use of chemicals may require considerably more knowledge, particularly of the biological effects of chemicals.

Environmental risk assessment (see Chapter 8) requires information on the properties of chemicals regarding their interactions with living organisms. It might not be necessary to know the properties with a very high degree of accuracy that can be provided by measurements in a laboratory, but it would be beneficial to know the properties with sufficient accuracy to make it possible to obtain a reliable quantification or develop an applicable ecotoxicological model for risk assessment and risk managements. As we apply safety factors, the required accuracy or the maximum acceptable standard deviation of the quantification and the ecotoxicological models is usually not very demanding. Therefore, estimation methods have been developed as an urgently needed alternative to measurements (see Chapter 4). These

TABLE 3.1

Overview of the Most Relevant Physical–Chemical, Environmental Properties of Toxic Compounds and Their Interpretation

Property	Interpretation
Water solubility	High water solubility corresponds to high mobility (an environmental disadvantage), but low bioaccumulation (an environmental advantage).
K_{ow} solubility ratio in octanol/water	High K_{ow} means that the compound is lipophilic. It implies that it has a high tendency to bioaccumulate and biomagnify and be sorbed to soil sludge and sediment. BCF and K_{oc} are well correlated with K_{ow}.
Biodegradability	This is a measure of how fast the compound decomposes into simpler molecules. A high biodegradation rate implies that the compound will not accumulate in the environment, while a low biodegradation rate may create environmental problems related to the increasing concentration in the environment and the possibilities of a synergistic effect with other compounds.
Volatilization, vapor	High rate of volatilization (high vapor pressure) implies that the pressure compound will cause an air pollution problem.
Henry's constant, He	He determines the distribution between the atmosphere and the hydrosphere.
pK	pH relative to pK determines whether the acid or the corresponding base is present. As the two forms have different properties, pH becomes important for the properties of the compounds.
Bioavailability	This concept expresses the quantity of the compound available for biodegradation. Bioavailability depends on a number of factors, including whether the compound is free/labile or bound and the ability of the compound to cross biological membranes.

Source: Jørgensen, S.E., *Pollution Abatement*, Elsevier, Amsterdam, 488pp., 2000.

are, to a great extent, based on the structure of the chemical compounds, the so-called QSAR and SAR methods (see Chapter 4), but it may also be possible to use allometric principles to transfer rates of interaction processes and concentration factors between a chemical and one or a few organisms to other organisms. Allometric principles are based on the size of the organism. Many rates are proportional to the surface area of the organism, as the surface area determines the interactions between organisms and their environment. Peters (1983) has published many allometric principles, including their applications of ecotoxicological parameters.

It may be interesting here to discuss the obvious question: Why is it sufficient to estimate the property of a chemical in an ecotoxicological context with 20%, or sometimes with as high as 50% or even higher, uncertainty? Ecotoxicological assessment usually gives an uncertainty of the same order of magnitude, which means that the indicated uncertainty may be sufficient from the modeling point of view, but can results with such an uncertainty be used at all? The answer is often "yes" because in most cases we want to assure ourselves that we are (very) far from a harmful or very harmful level. We use a safety factor of 10–1000 (most often 50–100) in the development of environmental risk assessment (ERA). When we are concerned with extremely harmful effects, such as the complete collapse of an ecosystem or a health risk for a large human population, we will inevitably select a safety factor, which is very high. In addition, our lack of knowledge about the synergistic effects and the presence of many compounds in the environment at the same time forces us to apply a very high safety factor. In such a context, we will usually opt for a concentration in the environment, which is of lower order of magnitude than that corresponding to a slightly harmful effect or considerably lower than the NEC (noneffect concentration). It is analogous to the construction of bridges by civil engineers. They make very sophisticated calculations (develop models) that take into account factors such as the speed of wind, snow, temperature changes, and so on and afterward they multiply the results by a safety factor of 2–3 to ensure that the bridge will not collapse. They use safety factors because the consequences of a bridge collapse are unacceptable.

The collapse of an ecosystem or a health risk to a large human population is also completely unacceptable. So, we should use safety factors in ecotoxicological quantification and modeling to account for the uncertainty. Owing to the complexity of the system, the simultaneous presence of many compounds, and our present knowledge, or rather lack of knowledge, we should, as indicated above, use a safety factor of 10–100 or sometimes even 1000. If we use safety factors that are too high, the risk is only that the environment will be less contaminated at maybe a higher cost. Besides, there are no alternatives to the use of safety factors. We can increase our ecotoxicological knowledge step by step, but it will take decades before it may be reflected in considerably lower safety factors. A measuring program of all processes and components is impossible due to the high complexity of the ecosystems. This does not imply that we should not use the information on the measured properties available today. The measured data will almost always be more accurate than the estimated data. Furthermore, the use of measured data within the network of estimation methods will improve the accuracy of the estimation methods. Fortunately, several handbooks on ecotoxicological parameters are available. References to the most important are given below:

Jørgensen, S.E., S.N. Nielsen, and L.A. Jørgensen. 1991. *Handbook of Ecological Parameters and Ecotoxicology*. Elsevier, Boca Raton, FL. Year 2000 published as a CD called Ecotox. It contains 3 times the amount of parameter in the 1991 book edition.

Howard, P.H. et al. 1991. *Handbook of Environmental Degradation Rates*. Lewis Publishers, Boca Raton.

Verschueren, K. Several editions have been published, the latest being in 2007. *Handbook of Environmental Data on Organic Chemicals*. Van Nostrand Reinhold, New York.

Mackay, D., W.Y. Shiu, and K.C. Ma. *Illustrated Handbook of Physical-Chemical Properties and Environmental Fate for Organic Chemicals*. Lewis Publishers, Boca Raton.

Volume I. *Mono-Aromatic Hydrocarbons. Chloro-Benzenes and PCBs*. 1991.

Volume II. *Polynuclear Aromatic Hydrocarbons, Polychlorinated Dioxins, and Dibenzofurans*. 1992.

Volume III. *Volatile Organic Chemicals*. 1992.

Jørgensen, S.E., H. Mahler, and B. Halling Sørensen. 1997. *Handbook of Estimation Methods in Environmental Chemistry and Ecotoxicology*. Lewis Publishers, Boca Raton.

Estimation methods for the physical–chemical properties of chemical compounds were already applied 40–60 years ago (see Jørgensen et al., 1997) as they were urgently needed in chemical engineering. They are to a great extent based on contributions to a focal property by molecular groups and the molecular weight: boiling point, melting point, and vapor pressure as a function of temperature are examples of properties that were frequently estimated in chemical engineering by these methods. In addition, a number of auxiliary properties results from these estimation methods, such as the critical data and the molecular volume. These properties may not have a direct application as ecotoxicological parameters in environmental risk assessment, but are used as intermediate parameters that may be used as a basis for an estimation of other parameters. Further details are presented in Chapter 4.

Biodegradability, as shown in Table 3.1, is a very important parameter, but in some cases, highly dependent on the concentration of microorganisms (for more details, see Jørgensen, 2000; Jørgensen et al., 2015). Therefore, it may be beneficial to indicate it as a rate coefficient relative to the biomass of the active microorganisms in units of milligrams per gram dry weight per 24 h.

In microbiological decomposition of xenobiotic compounds, an acclimatization period from a few days to 1–2 months should be foreseen before the optimum biodegradation rate can be achieved. We distinguish between primary and ultimate biodegradation. Primary biodegradation is any biologically induced transformation that changes the molecular integrity. Ultimate biodegradation is the biologically mediated conversion of an organic compound into an inorganic compound and products associated with complete and normal metabolic decomposition.

The biodegradation rate is expressed in a wide range of units:

1. As a first-order rate constant (1/24 h)
2. As half-life time (days or hours)
3. Milligrams per gram sludge per 24 h (mg/(g 24 h))
4. Milligrams per gram bacteria per 24 h (mg/(g 24 h))
5. Milliliter of substrate per bacterial cell per 24 h (mL/(24 h cells))
6. Milligram COD per gram biomass per 24 h (mg/(g 24 h))
7. Milliliter of substrate per gram of volatile solids inclusive of microorganisms (mL/(g 24 h))
8. BOD_x/BOD_8, that is, the biological oxygen demand (BOD) in x days compared with complete degradation (–), named the BOD_x-coefficient
9. BOD_x/COD, that is, the biological oxygen demand in x days compared with complete degradation, expressed by means of COD (–)

The biodegradation rate in water or soil is difficult to estimate because the number of microorganisms varies by several orders of magnitude from one type of aquatic ecosystem to the next and from one type of soil to the next.

It is furthermore essential to consider the bioavailability (see Table 3.1), which certainly adds to the complexity of the crucial question: how much of these toxic compounds will remain after x days, months, or years? It is noticeable that the amount adsorbed onto suspended matter or soil is in equilibrium with the dissolved concentration (usually in water) and is not directly available. The adsorption coefficient K_{oc} is, as indicated in Table 3.1, well correlated to K_{ow}. K_{oc} is, with the first approximation, the concentration in the adsorbent (often soil) divided by the concentration in the solution (often water), provided the carbon content of the adsorbent (often soil) is 100%. The adsorption coefficient for soil denoted K_{ac} is $K_{oc}*f$, where f is the fraction of carbon in the soil. However, a more detailed calculation applies Freundlich's or Langmuir's adsorption isotherm.

Adsorption is described by Freundlich's or Langmuir's adsorption isotherm, respectively (Palmeri et al., 2014), as follows:

$$\log a = A + k \log c \tag{3.1}$$

$$a = K * c/(k' + c) \tag{3.2}$$

where a is the concentration in the adsorbent and c is the dissolved concentration. If k in Equation 3.1 is 1.0 (it is often close to 1.0), we obtain that a/c is a constant—the adsorption coefficient. If k' is $\gg$ c, in Equation 3.2, we obtain that a/c is a constant as well. The graphs are shown in Section 7.8, where adsorption is included as a physical–chemical process.

It is clear from this brief overview of the properties of chemicals that it is necessary to have a wide knowledge of the physical–chemical properties of the toxic compounds to be able to determine or calculate the concentration of the emitted compounds in various environmental compartments and spheres. In addition, the compounds can be transformed to other compounds by chemical processes. Calculation of the concentrations resulting from chemical reactions is presented in Chapter 7. These calculations require that we should either know or determine the equilibrium constants for the most important chemical processes: acid–base, redox, complex formation, and precipitation. Due to the complexity, however, it is difficult in most cases to develop ecotoxicological models and to be able to determine which compounds we have in which concentrations in the various environmental compartments and spheres. Chapter 9 focuses on the development of ecotoxicological models to determine the predicted environmental concentration, denoted as PEC. This concept will be explained in more detail in Chapter 8 about the environmental risk assessment (ERA).

In conclusion, to assess the PEC a number of properties of the chemical compounds are needed to determine the effects of all the compounds that are present in the environment and their concentrations. Another question is to translate PEC to an effect. This would require knowledge of another wide spectrum of properties of the chemical (toxic) compounds considered. These properties have provided information on the relationship between concentrations of various chemical compounds and their biological–ecological–toxicological effects. As presented in Chapter 2, the concentration as f (time) plays an essential role, as the overall effect is time dependent. Section 3.2 gives a brief overview of these chemical–toxicological properties.

3.2 Properties (Parameters) Determining Ecotoxicological Effects

The most commonly applied toxicological parameters are summarized in the definitions in Table 3.2.

The ecotoxicological effects are often presumed to exist at the organism level, but the effect may be on all levels of the ecological hierarchy: molecular, cellular, organ, organism, population, ecosystem, landscape, regional, and ecosphere.

TABLE 3.2

Ecotoxicological Properties and Their Definitions

Parameter (Property)	Definition
LC_{50}	Lethal concentration: 50 indicates the percentage mortality; other mortality may be applied. Often, the duration of the experiment is also indicated; 48 or 96 h are usually applied.
LD_{50}	Lethal dose: 50 indicates the percentage mortality; other mortality may be applied. Often, the duration of the experiment is also indicated; 48 or 96 h are usually applied.
MAC	Maximum allowable concentration.
EC	Effect concentration. The effect is indicated by, for instance, no growth and also the percentage of organisms affected.
NC	Narcotic concentration effect. The percentage of test organisms affected is indicated.
HC	Hazardous concentration with indication of the percentage of test organisms affected.
NEC	Noneffect concentration.

LC and LD are of course presumed on the organism level, while EC is often, but not always, used on the molecular level (an enzymatic process is affected) or on the cell level (malfunctions of the cells). The effect may however also be on normal biological growth and reproduction.

There are several very harmful ecotoxicological effects that require particular attention:

- Toxic substances may destroy enzymes or alter them so that they function improperly. Heavy metals, cyanide, and many pesticides have this effect.

- Mutation, alteration of DNA, is a natural process, but xenobiotic substances in the environment may also cause mutations and thereby result in birth defects. Mutation takes place mainly by two different alterations of the DNA. The amino groups in the amino bases may be replaced by OH groups or by alkylation—most often by an attachment of a methyl group to an N atom in the amino bases.

- Cancer often results from alteration of cellular DNA, which may imply an uncontrolled replication and growth of the body's own cells. Most chemicals that are regarded as carcinogens require chemical activation to produce ultimate carcinogens. These chemicals are therefore considered procarcinogens. A few components are direct-acting carcinogens, for instance, benzo(*a*)pyrene (see Chapter 6). It should be emphasized that most cancers are not caused by carcinogenic agents but are natural processes in which there is a strong genetic component, which may have been enhanced due to a carcinogenic agent.

- The earlier the stage of development of any organism, the more vulnerable it is to damages to embryonic and fetal cells and to eggs or sperm cells, resulting in birth defects. Chemical substances causing such defects are called teratogens.

- The defense of our immune system against viral, bacterial, and protozoal infections destroys and neutralizes cancerous cells and resists the direct effect of xenobiotic toxicants. Chemicals may suppress the immune system, impairing the immune system's ability to fight diseases. An overstimulation of the immune system can, however, cause ill effects, known as allergy or hypersensitivity, which may be the result of a xenobiotic substance exposure.

- The endocrine system regulates metabolism and reproduction of organisms. Some toxic compounds can disrupt the function of the endocrine system, resulting in an abnormal behavior of the reproductive system, including dysfunction, alteration in secondary, sexual characteristics, and abnormal levels of blood steroids. Some toxic substances are hormonally active agents that exhibit hormone-like behavior. The most significant of these are estrogens that can imitate the function of the female sex hormone estrogen due to their similar structure to that of estrogen. They are often present in wastewater, but in small concentrations, which are, however, sufficient to alter the sexual characteristics of fish, frogs, and alligators.

Knowledge of the ecotoxicological effects, making it possible to translate the concentration in an environmental compartment, may be found by the use of the fugacity models presented in Chapter 2 or by the use of a more complex ecotoxicological model applied to develop an environmental impact assessment (EIA) (see Chapters 8 and 9). An ecotoxicological model may often include a translation of the concentration to an effect, as will be presented in more detail in Chapter 9.

References

Jørgensen, S.E. 2000. *Pollution Abatement*. Elsevier, Amsterdam, 488pp.

Jørgensen, S.E., B. Halling-Sørensen, and H. Mahler. 1997. *Handbook of Estimation Methods in Environmental Chemistry and Ecotoxicology*. CRC Press, Lewis, New York, 230pp.

Jørgensen, S.E., J.C. Marques, and S.N. Nielsen. 2015. *Integrated Environmental Management: A Transdisciplinary Approach*. CRC Press, Boca Raton, 380pp.

Palmeri, L., A. Barausse, and S.E. Jørgensen. 2014. *Ecological Processes Handbook*. CRC Press, Boca Raton, 386pp.

Peters, R.H. 1983. *The Ecological Implications of Body Size*. Cambridge University Press, Cambridge, 329pp.

4

Estimation of Ecotoxicological Parameters

4.1 Introduction

All of us would like to enjoy the benefits of using chemicals but cannot accept the harm they may cause. This conflict raises several urgent questions, which have been dealt with in the earlier chapters. These questions cannot be answered without models, and we cannot develop models without knowing the important parameters, at least within some ranges. We cannot perform an integrated environmental management without knowing the properties of the chemicals that we use in our everyday life and in the industries. In addition to the physical–chemical properties that were discussed in Chapter 3, we should know the effects on a wide range of organisms. It means that we should be able to obtain the LC_{50}- and LD_{50}-values, the MAC and NEC values (for the abbreviations and the definitions used, see Chapter 3) and the relationship between the various possible sublethal effects and concentrations, the influence of the chemical on fecundity and the carcinogenic and teratogenic properties. We should also know the effect on the ecosystem level. How do the chemicals affect populations and their development and interactions, that is, the entire biological network of the ecosystem? If we modestly require the effect of the toxic substances on 1000 organisms to represent the about 5 million species that Earth gives life to, and for each interaction between a chemical and an organism needs 50 parameters, we need to know totally $100{,}000 \times 1000 \times 50 = 5$ billion parameters. An optimistic estimation would be that we know 0.5% of what we ought to know about chemicals. We can make a proper integrated environmental management for about 500 chemicals and we are able to make reasonable environmental risk assessment for maybe 5000 chemicals. Owing to the very limited knowledge about chemicals, applied in the industrialized society, we need methods to estimate the parameters and properties that is not available in the vast chemical, ecotoxicological, and toxicological literature.

The biodegradability of chemicals is generally a very important aspect that one needs to know, as it indicates how long a chemical will remain in the environment before it is degraded. The biodegradation in waste treatment plants is often of particular interest, in which case the %BOD may be used.

It is defined as the 5-day BOD as percentage of the theoretical BOD. It may also be indicated as the BOD_5-fraction. For instance, a BOD_5-fraction of 0.7 will mean that BOD_5 corresponds to 70% of the theoretical BOD. It is also possible to find an indication of BOD_5 percentage removal in an activated sludge plant. Jørgensen et al. (1991, 2000) contain information about the biodegradability of many chemical compounds. See also Section 3.1, which contains an overview of biodegradation expressions.

4.2 Application of Correlation Equations

The water solubility, the partition coefficient octanol–water, K_{ow}, and Henry's constant are crucial parameters in the network of estimation methods because many other parameters are well correlated with these two parameters. These three properties can be found for a number of compounds, or be even estimated with reasonably high accuracy using knowledge of the chemical structure, that is, the number of various elements, the number of rings, and the number of functional groups. In addition, there is a good relationship between water solubility and K_{ow}; see Figure 4.1. Recently, many

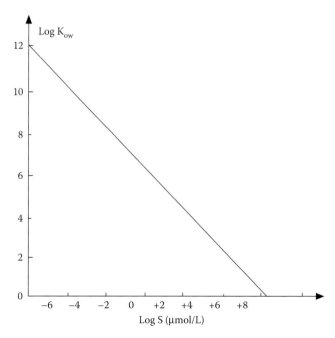

FIGURE 4.1
Log–log relationship between water solubility (unit: µmol/L) and octanol–water distribution coefficient.

good estimation methods for these three core properties have been developed and their applicability approved.

During the last couple of decades, several correlation equations have been developed based on a relationship between water solubility, K_{ow}, or Henry's constant on the one hand and physical, chemical, biological, and ecotoxicological parameters for chemical compounds on the other. The most important of these parameters are the adsorption isotherms of soil–water; the rate of the chemical and microbiological degradation processes—hydrolysis, photolysis, and chemical oxidation; the biological concentration factor (BCF); the ecological magnification factor (EMF); the uptake rate; the excretion rate; and a number of ecotoxicological parameters. Both the ratio of concentrations in the adsorbing phase (soil) and in water at equilibrium, K_a, and BCF, defined as the ratio of the concentration in an organism and in the medium (water for aquatic organisms) at steady state presuming that both the medium and the food are contaminated, may often be estimated with a relatively good accuracy from expressions based on log K_{ow}. K_a is obtained as $f*K_{oc}$, where f is the fraction of the dry matter in soil that is carbon and K_{oc} is the value for the adsorbing phase presuming that it consists of 100% C. Log K_{oc} is obtained by an estimation equation

$$\log K_{oc} = A + B \log K_{ow}$$

and BCF is estimated by the following equation:

$$\log BCF = a \log K_{ow} + b$$

Numerous expressions with different a, b, A, and B values have been published (see Jørgensen, 2000; Jørgensen et al., 1997). Two of these relationships are shown in Figure 4.2, for mussels and fish. The difference between the two correlation lines are due to the difference in the size of fish (about 20–30 cm long) and mussels (about 5 cm long) and can be explained by the allometric principle. The specific area of mussels should be about five times larger and therefore the BCF for mussels should be expected to be five times larger than the BCF for fish. It is in accordance with the difference between the two lines in Figure 4.2, as log 5 = 0.7.

4.3 Biodegradation

Models enlisting artificial intelligence have been used as a promising tool to estimate the important parameter—biodegradability of chemicals. However, a (very) rough, first estimation can be made on the basis of the molecular structure.

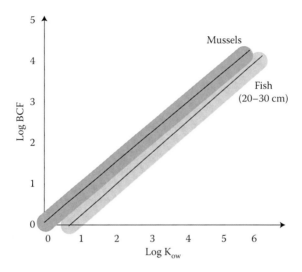

FIGURE 4.2
Two applicable relationships for octanol–water distribution coefficient and the biological concentration factor for fish and mussels. The shaded areas indicate the standard deviation and the difference between mussels and fish can be explained by the allometric principles.

The following rules can be used to set up these estimations:

1. Polymer compounds are generally less biodegradable than monomer compounds. 1 point for a molecular weight >500 and =1000, 2 points for a molecular weight >1000.

2. Aliphatic compounds are more biodegradable than aromatic compounds. 1 point for each aromatic ring.

3. Substitutions, especially with halogens and nitro groups, will decrease the biodegradability. 0.5 points for each substitution, although 1 point if it is a halogen or a nitro group.

4. Introduction of double or triple bonds will generally mean an increase in the biodegradability (double bonds in aromatic rings are of course not included in this rule). −1 point for each double or triple bond.

5. Oxygen and nitrogen bridges (–O– and –N– [or =]) in a molecule will decrease the biodegradability. 1 point for each oxygen or nitrogen bridge.

6. Branches (secondary or tertiary compounds) are generally less biodegradable than the corresponding primary compounds. 0.5 point for each branch.

Sum the total number of points and use the following classification:

1.5 points: The compound is readily biodegraded. More than 90% will be biodegraded in a biological treatment plant.

2.0–3.0 points: The compound is biodegradable. Probably about 10%–90% will be removed in a biological treatment plant. BOD_5 is 0.1–0.9 of the theoretical oxygen demand.

3.5–4.5 points: The compound is slowly biodegradable. Less than 10% will be removed in a biological treatment plant. $BOD_{10} = 0.1$ of the theoretical oxygen demand.

5.0–5.5 points: The compound is very slowly biodegradable. It will hardly be removed in a biological treatment plant and a 90% biodegradation in water or soil will take = 6 months.

= 6.0 points: The compound is refractory. The half-life time in soil or water is counted in years.

4.4 Estimation of Biological Parameters

Several useful methods for estimating biological properties are based on the similarity of chemical structures. The idea is that if we know the properties of one compound, the information may be used to find the properties of similar compounds. If, for instance, we know the properties of phenol, which is named the parent compound, the data may be used to give more accurate estimation of the properties of monochloro-phenol, dichloro-phenol, trichloro-phenol, and so on, and for the corresponding cresol compounds. Estimation approaches based on chemical similarity generally give more accurate estimation, but of course are also more cumbersome to apply, as they cannot be used generally in the sense that each estimation has a different starting point, namely, the parent compound, with known properties.

Allometric estimation methods presume (Peters, 1983) that there is a relationship between the value of a biological parameter and the size of the affected organism. These estimation methods are presented in detail in Jørgensen and Fath (2011).

The various estimation methods may be classified into two groups:

A. General estimation methods based on an equation of general validity for all types of compounds, although some of the constants may be dependent on the type of chemical compound or they may be calculated by adding contributions (increments) based on chemical groups and bonds.

B. Estimation methods valid for a specific class of chemical compounds, for instance, aromatic amines, phenols, aliphatic hydrocarbons, and so on. The property of at least one key compound is known. Based on the structural differences between the key compounds and all other compounds of the considered type—for instance, two chlorine atoms have substituted hydrogen in phenol to get 2,3-dichloro-phenol—and the correlation between the structural differences and the differences in the considered property, the properties for all compounds of the considered class can be found. These methods are therefore based on chemical similarity.

Methods of Class B are generally more accurate than that of Class A, but they are more cumbersome to use as it is necessary to find the right correlation for each chemical type, for instance, for phenols, for monochloro aromatic compounds, di-oxo-compounds, and so on. The correlations are therefore very specific. Furthermore, the requested properties should be known for at least one key component, which sometimes may be difficult when a series of properties are needed. If the estimation of the properties for a series of compounds belonging to the same chemical class is required, then it is tempting to use a suitable collection of Class B methods.

Methods of Class A form a network, which facilitates possibilities of linking the estimation methods together in a computer software system, like for instance estimation ecotoxicological parameters (EEP). It is available with this book at https://www.crcpress.com/product/isbn/9781498716529.

EEP contains many estimation methods, and detailed information about the use of the software is given in Section 4.5. The relationship between two properties in the applied network of estimation methods in EEP is based on the average result obtained from a number of different equations found in the literature. EEP is able to estimate several important properties of chemicals very fast, while, as mentioned above, it would require much more time to use the Class B methods. There is, however, a price for using such "easy-to-go" software. The accuracy of the estimations is not as satisfactory as with the more sophisticated methods based on similarity in chemical structure, but in many, particularly modeling, contexts, the results found by EEP can offer sufficient accuracy. In addition, under all circumstances it is always useful to come up with a first intermediate guess. In addition, EEP also gives the possibilities to estimate ecological and toxicological parameters by a limited number of Class B estimation methods, as will be illustrated in Section 4.5.

The software also makes it possible to start the estimations from the properties of the chemical compound already known. The accuracy of the estimation from the use of the software can be improved considerably by having knowledge about a few key parameters such as the boiling point, melting point, and Henry's constant, which can often be found in handbooks. It is furthermore possible to obtain software that is able to estimate Henry's constant and K_{ow} with generally higher accuracy than EEP. A combination of separate estimations of these two parameters prior to using EEP can be recommended. Another possibility would be to estimate a couple of key properties using chemical similarity methods and then use these estimations as known values in EEP. These methods for improving the accuracy will be discussed in Section 4.5. The network of EEP is illustrated in Figure 4.3. As pointed

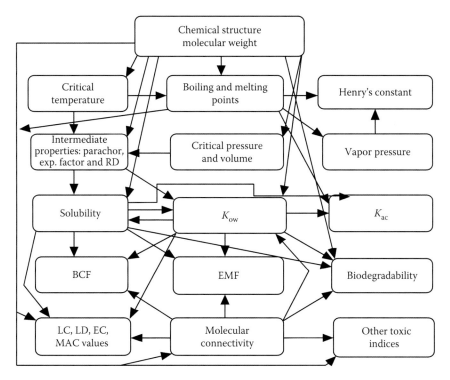

FIGURE 4.3
The network of estimation methods in EEP is shown. An arrow represents a relationship between two or more properties.

out above, it is a network of Class A methods. It should not be expected that the accuracy of the estimations is as high as possible to obtain by using the more specific Class B methods. By EEP, it is, however, possible to estimate the most pertinent properties directly and relatively from the structural formula, including the important property—biodegradability.

EEP is based on average values of results obtained by simultaneous use of several estimation methods for most of the parameters. It implies increased accuracy of the estimation, mainly because it gives a reasonable accuracy for a wider range of compounds. An arrow in Figure 4.3 represents, therefore, one or in many cases the average based on several equations. The property from where the arrow originates is used to calculate the property to which the arrow is going. If several methods are used in parallel, then a simple average of the parallel results have been used in some cases, while a weighted average is used in other cases where it has been found to be beneficial for the overall accuracy of the program. When parallel estimation methods give the highest accuracy for different classes of compounds, the use of weighting factors seems to offer a clear advantage.

It is generally recommended to apply as many estimation methods as possible for a given case study to increase the overall accuracy. Therefore, if the estimation by EEP can be supplemented by other applicable estimation methods, then it is strongly recommended to do so.

As mentioned earlier, we must also know the effects on a wide range of different organisms. It means that we should be able to find the LC_{50}- and LD_{50}-values, the MAC and NEC values, the relationship between the various possible sublethal effects and concentrations, the influence of the chemical on fecundity, and the carcinogenic and teratogenic properties. We should also know the effect on the ecosystem level. How do the chemicals affect populations and their development and interactions, that is, the entire network of the ecosystem? EEP offers the possibilities to estimate several of these parameters, partially by a limited number of Class B estimation methods, which are included in the software under the label of ecotoxicological estimations. The limitations in the use of these Class B methods are clearly indicated. It may be either a well-defined class of compounds or a well-defined organism or both.

4.5 Application of EEP

When EEP is downloaded from https://www.crcpress.com/product/isbn/9781498716529, the screen picture as shown in Figure 4.4 will emerge.

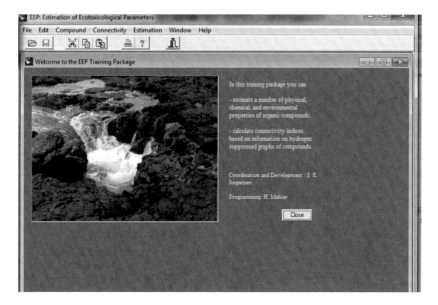

FIGURE 4.4
(**See color insert.**) The opening screen image of EEP.

Open now by use of file, the screen picture new/open compound. The screen picture shown in Figure 4.5 will appear. It is possible to reelect a chemical that has been applied for estimations previously. In this example used to illustrate how to apply the software, we will select 1-chloro-4-nitro-benzene as an illustrative example and by clicking OK, the screen picture of Figure 4.6 will emerge. As indicated in Figure 4.5, it is also possible to select a new compound and in that case the name of the chemical and the user must be informed. Figure 4.6 requires information of the number of elements of the considered compound. The elements number 1–92 are shown and in this case it is of course indicated that there are four hydrogen, six carbon, two oxygen, and one nitrogen atom for the selected chemical 1-chloro-4-nitro-benzene. When the number of elements has been indicated, click OK and Figure 4.7 will appear.

In Figure 4.7, it is necessary to indicate the molecular structure for the selected compound. For the selected example, it implies that it is required to indicate that the chlorine atom is in the ring and so are the six carbon atoms. Furthermore, it is needed to indicate that there is one nitro group and that the ring is aromatic. When the information about the molecular structure has been given, OK is clicked and the estimation of the physical–chemical parameters have been calculated based on the many equations

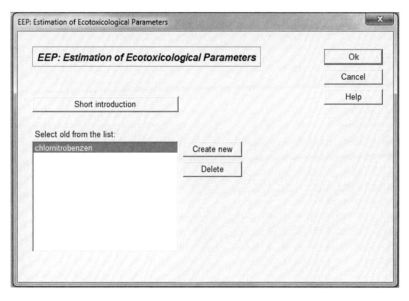

FIGURE 4.5
(**See color insert.**) Either a previously applied compound or a new compound can be selected. In the continuous illustration of the application of EEP, we will use 1-chloro-4-nitro-benzene. It means we will select from the list of already-used compounds. For a new compound, the chemical name and the initial of the user must be given. When the compound has been selected, click OK and Figure 4.6 appears.

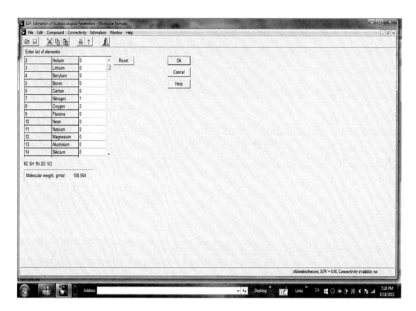

FIGURE 4.6
This screen image requires information of the number of elements of the considered compound. In this case, four hydrogen atoms, six carbon atoms, two nitrogen atoms, and one chlorine atom must be indicated.

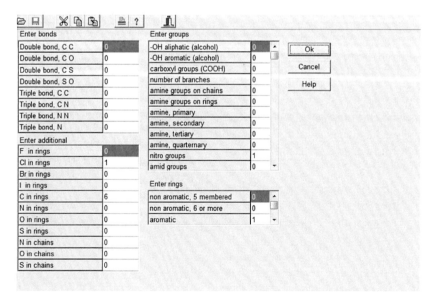

FIGURE 4.7
It is the fourth image of EEP. The information of the molecular structure has to be indicated. In this case, that the number of atoms in the ring is one chlorine atoms and six carbon atoms. Furthermore, it is indicated that the molecule contains one nitro group and one aromatic ring. After this information has been given, click OK.

applied in the network of EEP (see Figure 4.3). It is however, as mentioned above, possible to enter the information of some known parameters (properties) to increase the accuracy of the estimation. It is possible under estimation to go to "choose users own value." The screen image, Figure 4.8, will be shown. In this case, the boiling point and the melting point for 1-chloro-4-nitro-benzene have been applied. They are easily found in handbooks of physics and chemistry, and these two parameters can almost always be found and used to increase the accuracy or reduce the standard deviations of the estimations. Usually, it is beneficial to also apply known values of the solubility and of Henry's constant if it is possible to provide the information. It would in most cases imply a significant reduction of the standard deviation for the estimation.

It is now possible to provide the information of the physicochemical parameters and properties, including the biodegradability, using the information of the values of parameters or properties given in the screen image, see Figure 4.8, and the network of estimation methods for the other properties. The results of the estimations and all the physical–chemical properties, including users' own values, will be shown on the screen by opening under estimation "estimation of physicochemical parameters." The screen image is

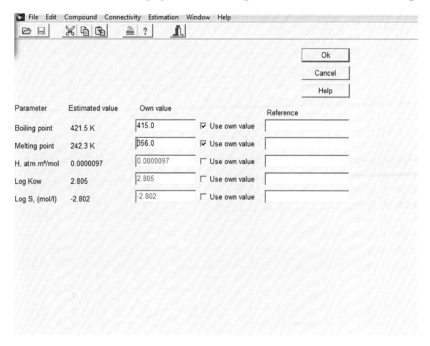

FIGURE 4.8
The boiling point and the melting point for 1-chloro-4-nitro-benzene have been applied in the shown case of application of users' own values. They are easily found in handbooks of physics and chemistry, and these two parameters can almost always be found and used to increase the accuracy or reduce the standard deviations of the estimations.

```
EEP: Estimation of Ecotoxicological Parameters

17-08-15

User name: sej
Summation of results for: chlornitrobenzen
(physicochemical parameters only)
------------------------------------------------------------------
Atomic composition: 6C 5H 1N 2O 1Cl
Mol weight                      : M = 158.56 g/mol
Temperature, actual             : T = 293.0 K
Pressure                        : P = 1.00 atm
------------------------------------------------------------------
Boiling point                   : Tb = 415.0 K (141.8°C)
                                  (user specified value)

Melting point                   : Tm = 356.0 K (82.8°C)
                                  (user specified value)

Critical temperature            : Tc = 1192.9 K (919.7°C)
Critical pressure               : Pc = 33.46 atm
Critical volume                 : Vc = 375.00 cm³/gram mol
Reduced temperature             : Tr = 0.25
Reduced pressure                : Pr = 0.030

Henrys constant                 : H = 0.000010 atm m³/mol
Vapour pressure                 : P = 0.0017 atm

Molal volume                    : Vm = 130.38 cm³/mol

Molar refraction                : RD = 36.55 cm³

Kow, octanol/water coefficient: Log Kow = 2.805
Solubility (water)              : Log S = -2.802 (mol/l)
                                  S = 0.0250 g/100 mL (20°C)
                                  S = 250.438 mg/L (20°C)
Adsorption isotherm             : log(Koc) = 2.209
Molecular diffusivity (air)     : Da = 0.0296 cm² /s
Molecular diffusivity (water) : Dw = 0.00000767 cm² /s

Biodegradability                : Biological half life = 1-10 months
------------------------------------------------------------------
```

FIGURE 4.9
The result of the estimation of physicochemical parameters, using the values in Figure 4.8.

shown in Figure 4.9. The results are only slightly different from the estimation without the two users' own values of the boiling point and the melting point. The critical temperature, the vapor pressure, the water solubility, and the adsorption isotherm are improved slightly by application of the right boiling point and melting point. Usually, it is possible to find more users' own values and it can strongly be recommended to try to find as many properties of the focal compound as possible to obtain the highest accuracy. By use of handbooks and databases, it is in most cases possible to find at least several values of parameters, determined by measurements and accessible in the scientific literature.

It can be recommended in additions to CRC Press *Handbooks of Physics and Chemistry* to at least have the following:

Howard, P.H. et al., 1991. *Handbook of Environmental Degradation Rates.* Lewis Publishers, Boca Raton.

Jørgensen, S.E., Nielsen, S.N., and Jørgensen, L.A. 1991. *Handbook of Ecological Parameters and Ecotoxicology*, Elsevier, Amsterdam, 1991.

Year 2000 published as a CD called Ecotox. It has Leif Jørgensen as first author and it contains 3 times the amount of parameter in the 1991 book edition.

Jørgensen, S.E., Mahler, H., and Halling Sørensen, B., 1997. *Handbook of Estimation Methods in Environmental Chemistry and Ecotoxicology.* Lewis Publishers, Boca Raton.

Mackay, D., Shiu W.Y., and Ma, K.C. *Illustrated Handbook of Physical–Chemical Properties and Environmental Fate for Organic Chemicals.* Lewis Publishers, Boca Raton.

Verschueren, K. Several editions have been published the latest in 2007. *Handbook of Environmental Data on Organic Chemicals.* Van Nostrand Reinhold, New York.

Volume I. *Mono-Aromatic Hydrocarbons. Chloro-Benzenes and PCBs.* 1991.

Volume II. *Polynuclear Aromatic Hydrocarbons, Polychlorinated Dioxins, and Dibenzofurans.* 1992.

Volume III. *Volatile Organic Chemicals.* 1992.

The estimation of ecological and toxicological parameters and properties by EEP are based on more specific estimation methods and they often require the use of connectivity, which can be calculated by the structural molecular formula. The estimations of ecological and ecotoxicological parameters are in most cases more accurate, when they are based on connectivity. Calculations of connectivity have therefore been included in EEP. Information about which atoms are connected to which atoms has to be given, as indicated in Figure 4.10. The procedure applied to obtain the connectivity by EEP is presented under the label of instructions:

1. Draw (on paper) a hydrogen-suppressed graph of the compound.
2. Select a numbering of all nonhydrogen atoms of the graph.
3. In the list (top left), change the order of the atoms using the red arrows to make it correspond to your numbering on paper.
4. Indicate all bonds of the compound in the connectivity matrix, according to your numbering.
5. Enter delta values and valence delta values of all atoms in the top left list (click on "?").
6. Click the Results button to see a report of the resulting connectivity indices.

The screen image will contain this six-step procedure.

The results of the connectivity determination are displayed in Figure 4.11. The results are applied in the further estimation of ecological and toxicological properties. These estimation results are shown in Figures 4.12 through 4.15.

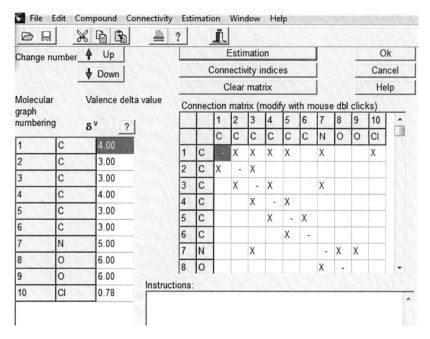

FIGURE 4.10
The screen image applied to calculate the connectivity, which is important for the estimation of ecological and toxicological parameters.

The methods are more specific with respect to both compound and organism, although estimations in a few cases of general physical–chemical parameters, based on connectivity, are also included. The group of compounds and the group of organisms have to be selected by the application of estimation methods.

Figure 4.12 gives the results of estimation of the inhibition by the chemical, 1-chloro-4-nitro-benzene, used as example, of aerobic decomposition by *Pseudomonas*. The results include an indication of how to use the information to calculate the changed Vmax values due to the inhibition, and the reference to the applied specific estimation method is given. For details, see Figure 4.12.

The use of general estimation methods with respect to compounds can be illustrated by the calculation of BCF using log K_{ow} for BCF of fish. Two sets of results are displayed and the average result of the two sets are shown too. Reference and information about standard deviation are included.

84 organic chemicals. Using $\log(K_{ow})$

Organism/environment: Fish (fathead minnow, bluegill sunfish, rainbow trout, mosquitofish)

Equation: $\log(BCF) = 0.76 * \log(K_{ow}) - 0.23$

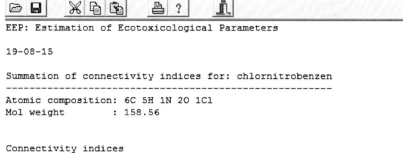

```
EEP: Estimation of Ecotoxicological Parameters

19-08-15

Summation of connectivity indices for: chlornitrobenzen
--------------------------------------------------------
Atomic composition: 6C 5H 1N 2O 1Cl
Mol weight        : 158.56


Connectivity indices

Order              0      1      2      3      4      5      6
--------------------------------------------------------------
Path             7.684  4.733  5.114  4.306  2.720  1.611  0.518
Cluster                        1.640  0.204  0.000  0.000
Path/Cluster                          3.954  2.520  1.601
Chain                          0.000  0.000  0.000  0.000
--------------------------------------------------------------


Valence Connectivity indexes

Order              0      1      2      3      4      5      6
--------------------------------------------------------------
Path             5.705  3.248  3.131  2.396  1.252  0.510  0.125
Cluster                        0.649  0.059  0.000  0.000
Path/Cluster                          1.412  0.665  0.270
Chain                          0.000  0.000  0.000  0.000
--------------------------------------------------------------
```

FIGURE 4.11
The result of the connectivity calculations are shown. These results are applied in the estimation of ecological and ecotoxicological properties. How the connectivity indices are applied are not shown but can be found in the references given after the estimation results.

Result: Log(BCF) = 1.90

Reference: *Handbook of Chemical Property Estimation Methods*, 1990. P. 5–5

36 organic chemicals. Using log(S)

Organism/environment: Fish (brook trout, fathead minnow, bluegill sunfish, rainbow trout, carp)

Equation: Log(BCF) = 2.791 – 0.564*log(S)

Result: Log(BCF) = 4.37

Reference: *Handbook of Chemical Property Estimation Methods*, 1990. P. 5–5

All results: mean log BCF = 2.10, s.d. 1.17, n = 35

Figure 4.13 shows the estimation result of log(K_{oc}) for halogenated aromatics for a particular yellow-brown soil, which has a high humic acid and iron content.

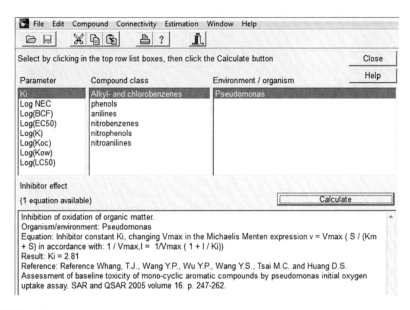

FIGURE 4.12
The inhibition of organic matter oxidation by *Pseudomonas* is estimated. The screen image shows how to apply the results and the reference where further details can be found.

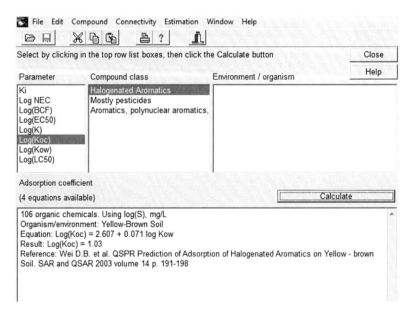

FIGURE 4.13
The estimation result of log K_{oc} for halogenated aromatics is shown for yellow-brown soil.

Figure 4.14 illustrates the result of estimation log K_{ow} for substituted benzenes using connectivity. The estimation gives as result log $K_{ow} = 3.51$, which can be compared with the estimation result given in Figure 4.9, where it is indicated that log $K_{ow} = 2.805$. The difference is a factor 5, which of course is high, but on the other side is the log value applied in many cases. Moreover, the estimation methods are not very accurate if they are used solely. In this case it was not possible to find the very accurate measured value, but generally it is recommended to apply an average of several estimation methods under all circumstances, where it is possible. Probably a value of log $K_{ow} = 3.15$ (average of 3.51 and 2.805) could be applied.

Figure 4.15 illustrates the results of estimating a toxicological parameter, LC-50, for mice. The standard deviation is high for all toxicological estimations. More combinations of compounds and organisms can be found by EEP, also for the illustrative example 1-chloro-4-nitro-benzene, but the selected five estimations are representative for the possibilities and the limitations that EEP offers.

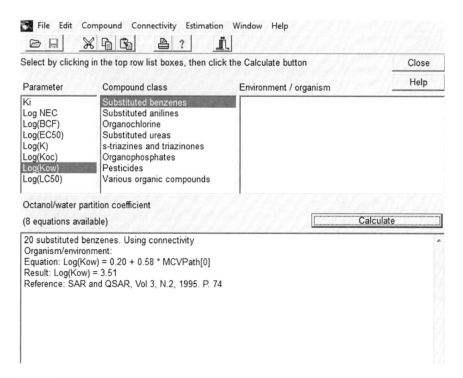

FIGURE 4.14
The result of estimating log K_{ow} using the connectivity is shown.

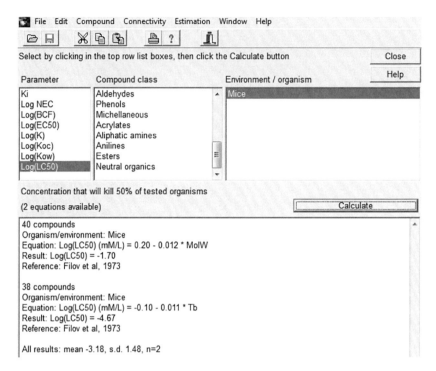

FIGURE 4.15
Filov et al. (1973) is the reference for these two sets of estimation of log LC$_{50}$. The two estimations are very different and generally have the toxicological estimations—the very specific methods, too—a very high standard deviation.

4.6 Conclusions

Estimation methods are of course not as accurate as measurements but due to the limited number of measurements and determinations that are available, we are highly dependent on estimation methods. The general estimation methods are easier and faster to use, but will in most cases imply a higher standard deviation than the more specific methods. EEP, which is mainly a general estimation tool, can estimate the most applied physical–chemical parameters very fast, in addition to some ecological and ecotoxicological properties. EEP is recommendable as a teaching tool in ecotoxicology and environmental chemistry because it is easy and fast to use; but if an estimation is important, for instance, in ERA, it would be advisable to use several estimation methods and use the average or the weighted average of the results. There are in most cases a wide spectrum of estimation methods available in the literature, which are rather easy to find, and their application is clearly explained. It will require of course much more time than just to use EEP, but it will under all circumstances most probably be much faster than setting up laboratory measurements of the properties. As the development of ERA for the many chemicals that we are using—sometimes almost blindly—is crucial for our environment, it is very fortunate that we have estimation methods.

___

References

Filov et al. 1973. *Quantitative Toxicology.* John Wiley & Sons, New York, 320pp.

Jørgensen, S.E. 2000. *Pollution Abatement.* Elsevier, Amsterdam, 488pp.

Jørgensen, S.E. and B. Fath. 2011. *Fundamentals of Ecological Modelling.* 4th edition. Elsevier, Amsterdam, 400pp.

Jørgensen, L.A., S.E. Jørgensen, and S.N. Nielsen. 2000. *Ecotox.* CD. Elsevier, Amsterdam.

Jørgensen, S.E., H. Mahler, and B. Halling Sørensen. 1997. *Handbook of Estimation Methods in Environmental Chemistry and Ecotoxicology.* Lewis Publishers, Boca Raton.

Jørgensen, S.E., S.N. Nielsen, and L.A. Jørgensen. 1991. *Handbook of Ecological Parameters and Ecotoxicology.* Elsevier, Amsterdam, 1991. Year 2000 published as a CD called Ecotox. It has Leif Jørgensen as first author and it contains 3 times the amount of parameter in the 1991 book edition.

Peters, R.H. 1983. *The Ecological Implications of Body Size.* Cambridge University Press. Cambridge, 329pp.

5

Global Element Cycles

5.1 The Biochemistry of Nature

The biochemistry for primitive cells and the most advanced animals, the mammals, are surprisingly similar. The metabolism follows approximately the same pathways. The core processes for photosynthesis are also the same for all plants. It is therefore not surprising that the elementary composition of various organisms is very similar.

The wet weight composition may be different because the water content varies more than the dry matter composition from organism to organism. A typical dry matter content of plants may be 12.5%. The composition provided in Table 5.1 is based on good approximations applicable to all plants and even to animals (Jørgensen, 2000).

Table 5.1 lists the 19 elements that are generally found in all organisms. Slightly more than a handful of elements can be found in the biosphere in addition to the 19 listed at low concentrations, for instance, iodine and flourine. A few trace elements have been found to be characteristic in just a few species: selenium, nickel, vanadium, chromium, and even the toxic element cadmium may substitute into enzymes that typically use zinc. In principle, all the applied elements can be a limiting factor for growth. A limiting element must, of course, be absolutely necessary for the considered organism and be present in a concentration that may be lower than the other important elements relative to its use for building new biomass for the focal organism. The composition of the ecosphere reflects, however, the composition of the organisms, which means that the elements are often present in a ratio close to concentrations as in Table 5.1. Furthermore, the composition of organisms is not an exact unchangeable value but rather a range. For instance, most plants can manage to grow with a phosphorus content of 0.4%–2.0%. If the environment has a high phosphorus concentration relative to the need of the organisms, the organism will accumulate more phosphorus, and if the phosphorus concentration of the environment is low, the plants are able to adapt to the conditions and cope with less phosphorus up to a lower limit of course. This adaptability of the plants to the composition of the environment is reflected in the range of concentrations in the organisms and implies that a

TABLE 5.1

Average Elemental Composition of Freshwater Plants (Dry-Weight Basis)

Element	Plant Content (%)
Oxygen	11.1
Hydrogen	1.4
Carbon	45.5
Silicon	9.1
Nitrogen	4.9
Calcium	2.8
Potassium	2.1
Phosphorus	0.56
Magnesium	0.49
Sulfur	0.42
Chlorine	0.42
Sodium	0.28
Iron	0.14
Boron	0.007
Manganese	0.0049
Zinc	0.0021
Copper	0.0007
Molybdenum	0.00035
Cobalt	0.000014

limiting concentration of an element is not a sharp exact value, but is a range that gradually at lower concentrations will limit growth more and more.

Table 5.2 provides the approximate ranges for C, N, and P for different organisms to show (1) the variability among species and (2) the possible ranges. The differences among species are not very pronounced, while the ranges may be relatively wide for phosphorus, which they generally are for elements with low biochemical concentrations.

TABLE 5.2

Approximate Ranges of C, N, and P Concentrations as % Dry Matter for Different Organisms

Organisms	%C	%N	%P
Terrestrial plants	36–64	0.3–6.4	0.02–1.0
Benthic invertebrates	35–57	6–12	0.2–1.8
Terrestrial insects	36–61	7–12.5	0.5–2.5
Birds and mammals	32–60	6–12	0.7–3.7
Fish	38–52	7–12	1.5–4.5
Zooplankton	35–60	7–12.5	0.5–2.5
Phytoplankton	35–60	5–12	0.5–2.5

5.2 Element Cycles and Recycling in Nature

Nature of course fully follows the conservation principle: conservation of matter and energy. All growth would stop unless the elements important to life recycles, which is nature's method to "surpass" the conservation of matter (Jørgensen, 2012). It is therefore of utmost importance that we do not create an imbalance of the global cycles due to discharge of pollutants to the spheres. Toxic substances may participate in some of the processes that make up the global cycles.

Renewable resources are used by man at a rate that in most cases is higher than the rate at which the resources are regenerated. This fact underscores that one of the most important sustainability criteria as mentioned in World Commission on Environment and Development (1987) is violated and that the renewable resources are decreasing. The nonrenewable resources are used by man at a rate that is higher than the rate at which alternatives to the nonrenewable resources are found. The decreasing renewable and nonrenewable resources demonstrate that Earth is not in sustainable development. Sustainable development is in this context used in the same sense as in the so-called Brundtland Report (World Commission on Environment and Development, 1987): A sustainable development means that we will hand over Earth with the same possibilities for the next generation to plan and live their life as the previous generations have given us when Earth was handed over to our generation.

Ecosystems do not use nonrenewable resources and completely recycle the elements that are applied in it for organisms. It does not imply that an ecosystem will always have resources that are needed for growth, as it has been discussed above. However, ecosystems will in these cases adjust the rate of consumption to what is possible on a long-term basis, while the recycling will continue.

Some nonrenewable resources for instance, iron, are recycled by man, but a 100% recycling is not possible and therefore nonrenewable resources are declining, although at a lower rate corresponding to the recycling.

The above 20–25 elements used by organisms of ecosystems are recycled in the ecosystems and the six elements that are considered absolutely necessary for life on Earth—C, H, O, N, P, and S—are of course particularly important for nature to recycle. Many ecological models focus on the element cycles to be able to describe properly the conditions of the ecosystems, as many of the main processes participate in the recycling processes (see Chapter 9 and Jørgensen and Fath, 2011). Figure 5.1 gives two examples—phosphorus and nitrogen cycles in a lake. The two examples include the inorganic form: either inorganic reactive phosphorus or ammonium and nitrate, the phosphorus and nitrogen in the phytoplankton, in zooplankton, in fish, in sediment, and in detritus. The phosphorus cycle includes the phosphorus in the pore water of the sediment. The detritus is mineralized to close the cycle, but the sediment can also release inorganic forms of phosphorus as phosphate

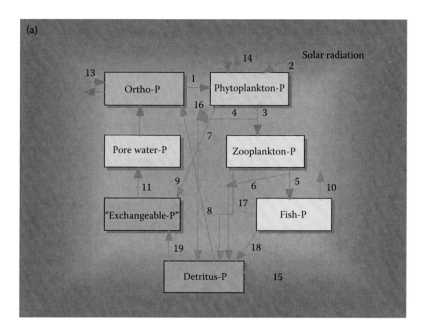

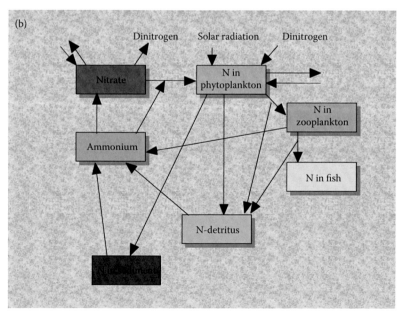

FIGURE 5.1
(**See color insert.**) Phosphorus (a) and nitrogen (b) cycles in a lake by using a simple food chain nutrient–phytoplankton–zooplankton–fish to describe the processes of the cycle. The cycles are closed by mineralization of detritus and by release of nutrient by the sediment. (Reprinted from *Fundamentals of Ecological Modelling*, 3, Jørgensen, S.E. and G. Bendoricchio, 620 pp., Copyright (2001), with permission from Elsevier.)

and nitrogen as ammonium. The cycles can include more or less details. As shown in Figure 5.1, the food chain is represented by nutrients, phytoplankton, zooplankton, and fish only. Phytoplankton could be represented by different groups of phytoplankton as, for instance, nitrogen-fixing species, diatoms, and so on. The possible nitrogen fixing is indicated in the diagram and so are the possibilities for denitrification. The two diagrams for the phosphorus and nitrogen cycles are almost parallel. The process numbers on the phosphorus refer to the processes that are the same for the two cycles. The following processes are included: (1) uptake of nutrient, (2) photosynthesis (solar radiation converted to organic matter in phytoplankton), (3) grazing, (4) loss of feces to detritus by the grazing process, (5) predation of zooplankton by fish, (6) loss of feces to detritus by predation, (7) settling of feces, (8) mineralization of detritus, (9) and (19) settling of phytoplankton and detritus, (10) fishery, (11) mineralization taking place in the sediment, (12) diffusion, (13)–(15) exchange of reactive phosphorus, phytoplankton, and detritus with the environment (inflows and outflows), and (16)–(18) mortality of phytoplankton, zooplankton, and fish. Note that the nitrogen cycle includes nitrification, which is the oxidation of ammonium to nitrate.

The recycling in ecosystems, illustrated by the recycling of nitrogen and phosphorus in a lake in Figure 5.1, ensures that renewable resources, elements, can actually be renewed. Thus growth and evolution take place continuously in the ecosystems. The elements that are nonrenewable for the society, for instance, many metals, undergo chemical processes in nature and they may also to a certain extent be recycled, but in most cases far too slow and uncontrollable to satisfy human consumption and ecosystems are not dependent on the resources, that are nonrenewable for the society. Figure 5.2 illustrates how matter cycles with the solar radiation as energy source (Jørgensen, 2012).

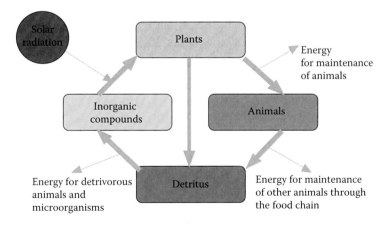

FIGURE 5.2
(**See color insert.**) An ecosystem is a biochemical reactor that recycles matter. The input of energy is delivered by the solar radiation. The biologically important elements cycle and carry the energy that is utilized by heterotrophic organisms to support life processes.

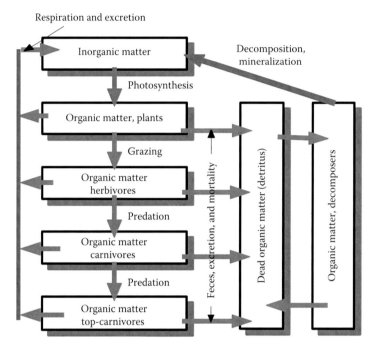

FIGURE 5.3
Biochemical recycling of matter in ecosystems. (Adapted from Jørgensen, S.E., *Pollution Abatement*, Elsevier, Amsterdam, 2000, 488pp.)

Figure 5.3 shows the resulting biochemical reactions of an ecosystem in more detail, that is, how the ecosystem works as a biochemical reactor, which, by use of the food chain, recycles matter. The biologically important elements are recycled and used again and again to build up important biochemical compounds such as proteins, lipids, and carbohydrates. These compounds carry the energy of the solar radiation and thereby support the maintenance of life and the cycling processes. The cycle may be compared with a Carnot cycle. The hot reservoir (the sun) delivers the energy, which is utilized to do work. The heat energy is delivered to the cold reservoir at ambient temperature (the temperature of Earth), becoming work energy after it has been used transformed to heat, which is delivered to the environment.

5.3 The Global Element Cycles

As ecosystems are open systems, the chemical and biochemical processes of ecosystems are interlinked and together form global element cycles. Figure 5.4 shows the global carbon cycle. The processes can be combined and

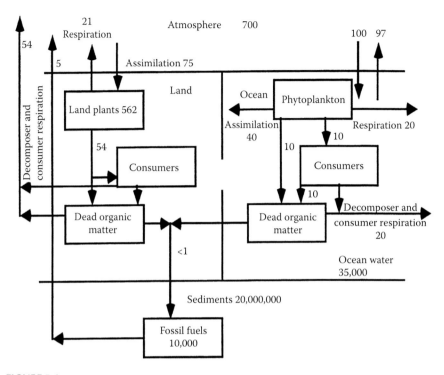

FIGURE 5.4
Global carbon cycle. Compartments are indicated as Gtons C and the flows as Gtons C/year. The imbalance is 5 Gtons C/year increasing relatively 2%–3% per year the last many years. Fortunately, about 60% of the emitted carbon-containing greenhouse gases will be taken up by the oceans. (Adapted from Jørgensen, S.E., *Pollution Abatement*, Elsevier, Amsterdam, 2000, 488pp.)

quantified by the use of different conceptual diagrams. The compartments are indicated as Gtons C and the flows as Gtons C/year. The global carbon cycle is imbalanced as about 5 Gtons/year is transferred from fossil fuel to the atmosphere of which 60% or 3 Gtons roughly is taken up by the oceans. The emission rate of carbon mainly in the form of carbon dioxide is furthermore increasing 2%–3% per year due to the growing demand for fossil fuel, that is, for work energy.

The imbalance of the global carbon cycle can only be solved effectively by cutting the consumption of fossil fuel by the development of alternative renewable energy sources such as wind and solar energy. The carbon dioxide concentration in the atmosphere has increased from about 280 ppm 100 years ago to about 400 ppm today (year 2014) due to the steadily increasing use of fossil fuel as energy source. Figure 5.5 shows the increase of carbon dioxide concentration in the atmosphere. The plot assumes that we do not change the trends of an exponential increase in the use of fossil fuel in the coming decades. The carbon dioxide concentrations up to the year 2014 are the measured values—it is seen that the concentration in 2014 is about

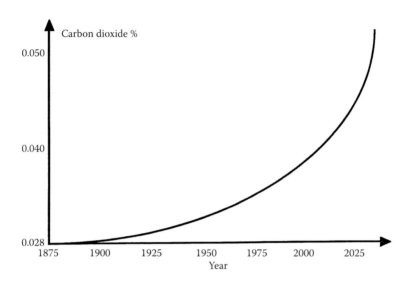

FIGURE 5.5
Carbon dioxide concentration plotted versus year, considering that 60% of the input is dissolved in the sea and that the consumption of fossil fuel continues to increase exponentially by 2%–3% per year.

400 ppm or 0.04%. A concentration of 500 ppm can be expected in about 14 years, before 2030, if the fossil fuel consumption continues to increase by the mentioned 2%–3% per year the coming years.

The global nitrogen balance is shown in Figure 5.6. The values in the compartments are in Gtons N and the fluxes are in Mtons N/year. It can easily be seen that the imbalance in the global recycling of nitrogen is 30 Mtons/year corresponding to the industrial fixation of nitrogen—it is the industrial production of nitrogen fertilizer. Furthermore, it is shown that this amount of nitrogen via river runoff reaches the oceans. Most of this nitrogen is accumulated in the sediment, where enormous amounts of nitrogen are and can still be accumulated. The main effect of the 30 Mtons N/year is not the creation of an imbalance in the oceans, but the effect of the elevated concentration in the coastal zones, where it causes an undesired eutrophication before it reaches the open sea.

The imbalance of the nitrogen cycle can be solved by either a decreased production of fertilizers, which is hardly possible in the world today with an increasing population, or by preventing the nitrogen to reach the hydrosphere by treatment of wastewater and drainage water. The use of constructed or natural wetland ecosystems offers here a good moderate-cost solution, particularly to treat drainage water, while technologically feasible methods are available for the treatment of municipal and industrial wastewater. The nitrate and ammonium are by these treatment methods transformed into dinitrogen, which is transported to the atmosphere. A pattern

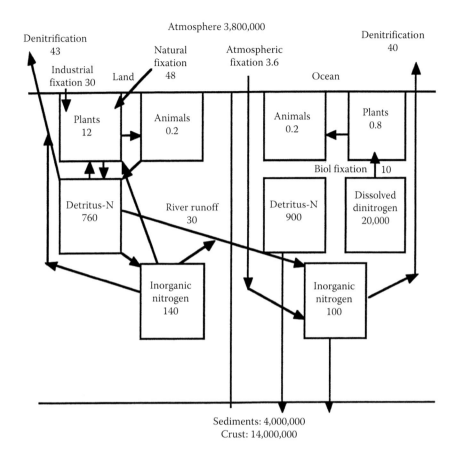

FIGURE 5.6
Global nitrogen cycle. The values in the compartments are in Gtons N and the fluxes are in Mtons N/year. The amounts and the annual transfers are only estimated within broad limits, except the amounts of nitrogen in the atmosphere, the natural nitrogen fixation, the industrial nitrogen fixation, and the denitrification. It can be seen that the imbalance in the global recycling of nitrogen is 30 Mtons/year corresponding to the industrial fixation of nitrogen and that this amount of nitrogen via river runoff reaches the oceans. Most of this nitrogen is accumulated in the sediment, where enormous amounts of nitrogen are accumulated. The main effect of the imbalance is on the coastal zones, where 30 Mtons/year cause an undesired eutrophication before it reaches the open sea.

of different types of wetlands in the landscape can reduce considerably the nitrogen loss to the hydrosphere.

The global sulfur cycle is shown in Figure 5.7. The transfer processes are shown in Mtons S/year. The river runoff transports sulfur from the land to the sea, but the sea transfers via the atmosphere by sea spray and biological decomposition a similar amount back to the land (the lithosphere). The most significant anthropogenic flux is the sulfur dioxide pollution of the atmosphere mainly due to the consumption of sulfur-containing fossil fuel.

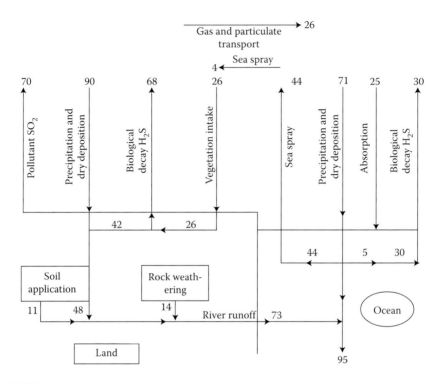

FIGURE 5.7
Transfer processes of sulfur shown in Mtons S/year. The river runoff transports sulfur from the land to the sea, but the sea transfers via the atmosphere by sea spray and biological decomposition a similar amount back to the land (the lithosphere). The most significant anthropogenic flux is the sulfur dioxide pollution of the atmosphere.

This flux exemplifies that toxic substance may participate in global natural cycles and thereby cause an imbalance and at the same time a toxic effect—in this case in the atmosphere.

The phosphorus cycle is also in imbalance. Phosphorus minerals are mined for the production of phosphorus fertilizers, applied in the lithosphere, but the major part is transported to the hydrosphere. As the amounts of phosphorous minerals are limited, it is necessary to regain and recycle more of the phosphorus that we are applying mainly as fertilizer.

All imbalances of global cycles or ecosystem cycles require a solution that respects the conservation principles. Matter or energy cannot be destroyed (or created) but only transformed, which implies that the energy sources used should be replaced by other energy forms that do not create imbalances of the spheres. It further implies that fossil fuel that causes accumulation of carbon dioxide in the atmosphere should be replaced by other energy forms and nitrogen in wastewater and drainage water should be transformed to dinitrogen, which would be harmless in the atmosphere, as it contains 78% dinitrogen.

TABLE 5.3

Atomic Composition of the Four Spheres

Element	Atoms % in			
	Biosphere	Lithosphere	Hydrosphere	Atmosphere
H	49.8	2.92	66.4	v.l.
O	24.9	60.4	33	21
C	24.9	0.16	0.0014	0.04
N	0.27	v.l.	v.l.	78
Ca	0.073	1.88	0.006	v.l.
K	0.046	1.37	0.006	v.l.
Si	0.033	20.5	Low	v.l.
Mg	0.031	1.77	0.034	v.l.
P	0.030	0.08	v.l.	v.l.
S	0.017	0.04	0.017	v.l.
Al	0.016	6.2	v.l.	v.l.
Na	Low	2.49	0.28	v.l.
Fe	v.l.	1.90	Low	v.l.
Ti	v.l	0.27	v.l.	v.l.
Cl	v.l.	v.l.	0.33	v.l.
B	v.l.	v.l.	0.0002	v.l.
Ar	v.l.	v.l.	v.l.	0.93
Ne	v.l.	v.l.	v.l.	0.0018

v.l. = very low.

Table 5.3 provides the composition of the spheres, which should be maintained on approximately the same level in the future to avoid imbalances of the global cycles and to the best possible compositions of the spheres for life.

All the elements cycle, but our knowledge is limited to an approximate quantification of the global cycles for the following elements: C, N, O, S, P, Fe, and Ca. However, the toxic elements cycle too. Mercury and lead are released into the atmosphere and are deposited in remote locations. The concentrations of mercury and lead are therefore significantly higher in the ice deposited during the last 100 years due to increased industrial activity.

References

Jørgensen, S.E. 2000. *Pollution Abatement*. Elsevier, Amsterdam, 488pp.

Jørgensen, S.E. 2012. *Introduction to Systems Ecology*. CRC Press, Boca Raton. Chinese edition 2013, 320pp.

Jørgensen, S.E. and G. Bendoricchio. 2001. *Fundamentals of Ecological Modelling*, 3rd edition. Elsevier, Amsterdam, 620pp.

Jørgensen, S.E. and B. Fath. 2011. *Fundamentals of Ecological Modelling.* 4th edition. Elsevier, Amsterdam, 400pp.

Jørgensen, S.E., J.C. Marques, and S.N. Nielsen. 2015. *Integrated Environmental Management: A Transdisciplinary Approach.* CRC Press, Boca Raton, 380pp.

World Commission on Environment and Development. 1987. *Our Common Future.* Oxford University Press, 360pp.

6

Toxic Substances in the Environment

6.1 Inorganic Compounds and Their Effects

The periodic table of today has 118 elements, but the last more than 20 elements are extremely unstable and have therefore no environmental interest with the exception of their radioactivity. Ninety-one of the elements are naturally occurring, namely, the first 92 elements from hydrogen to uranium except element number 43, technetium. These 91 elements can be classified into four groups (the numbers before the elements indicate the number in the periodic table):

A. Elements that are present in biological material in relative high concentrations: (1) hydrogen, (6) carbon, (7) nitrogen, (8) oxygen, (11) sodium, (12) magnesium, (14) silica, (15) phosphorus, (16) sulfur, (17) chlorine, (18) potassium, (19) calcium, and (26) iron.

B. Elements that are present in trace amounts in biological material. They may often be present in biological material in small concentrations: (5) boron, (9) fluorine, (24) chromium, (25) manganese, (27) cobalt, (28) nickel, (29) copper, (30) zinc, (34) selenium, (42) molybdenum, and (53) iodine. Too high concentrations may, however, cause environmental problems.

C. Elements that have ecotoxicological effect. They are harmful even in very small concentration and are therefore the elements that we have to consider in environmental management as ecotoxicological elements and compounds. They are the inorganic compounds that may cause pollution problems associated with a toxic effect. They are (13) aluminum, although there are many harmless aluminum compounds in nature, for instance, clay minerals, (33) arsenic, (38) strontium, (47) silver, (48) cadmium, (50) tin, (51) antimony, (52) tellurium, (56) barium, (80) mercury, (81) thallium, (82) lead, (83) bismuth, (84) polonium, (86) radon, (88) radium, (90) thorium, (92) uranium, and (94) plutonium. The last six elements are, in addition to being toxic, also radioactive. They are normally present in small concentrations and the main concern is their radioactivity. As five elements belonging to group B may give rise to ecotoxicological/environmental concerns due to the frequent occurrence of harmful environmental concentrations, namely, chromium, copper, nickel, selenium, and zinc, the overview in Section 6.2 will include these five elements and encompass 18 elements in all.

D. All other elements have no or only minor environmental interest due to their occurrence in low concentrations or low biological effects.

Table 6.1 gives the composition of average freshwater plants on wet basis. The table presents 19 elements out of the 24 elements that are included in groups A and B. The following elements are not included in Table 6.1: fluorine, chromium, nickel, selenium, and iodine, because they are present in very small concentration in freshwater plants—below 0.000002%. Compare this table with Table 5.1, which is based on the use of dry weight. The carbon is here 45.5/6.5 = 7 times higher, corresponding to a water content of 85.7% and a dry weight content of 14.3% in average.

The elements in groups B and C show bioaccumulation and biomagnifications, which are two important processes to be considered in all ecotoxicological evaluations (see Chapter 2).

TABLE 6.1

Average Freshwater Plant Composition on Wet Basis

Element	Plant Content (%)
Oxygen	80.5
Hydrogen	9.7
Carbon	6.5
Silica	1.3
Nitrogen	0.7
Calcium	0.4
Potassium	0.3
Phosphorus	0.08
Magnesium	0.07
Sulfur	0.06
Chlorine	0.06
Sodium	0.04
Iron	0.02
Boron	0.001
Manganese	0.0007
Zinc	0.0003
Copper	0.0001
Molybdenum	0.00005
Cobalt	0.000002

Source: Wetzel, R.G., *Limnology*, Second edition. Saunders College Publishing, New York, 828pp., 1983.

The 24 elements in classes A and B encompass the elements that are needed for the entire biosphere. All 24 elements are not necessarily essential for all organisms, but each organism will require most of these 24 elements for their growth. These elements are threshold agents (see Jørgensen, 2000 and Chapter 2). They are needed in a certain concentration for the growth of the organisms, but if they are present in a too high concentration, they may be harmful to the environment. A typical example is the eutrophication problem, which is a result of a too high concentration of nutrients (see Figure 5.1), but the nutrients (with emphasis on nitrogen, carbon, phosphorus, and sulfur) are absolutely necessary for all living organisms. Below, we have included five class B elements in the list of elements as mentioned above, because they are frequently present in concentrations in the environment that may cause ecotoxicological effects. Practically, nonthreshold agents or gradual agents in any concentration are harmful for the environment. The harmful effect (group C elements) may be proportional to the concentration or there may be another relationship between the effect and the concentration; for instance, the harmful effect is proportional to the concentration in exponent 2.

Pollution by many metals is an important soil pollution problem. The expression "heavy metal pollution" is frequently used to cover the problem of many metals. In principle, heavy metals are metals with a specific gravity equal to or greater than iron, but, for instance, cadmium is often included as a heavy metal, although the specific gravity is less than iron, because it is a metal and the environmental problems are severe and similar to the problems caused by many heavy metals. Three of the C elements require particular environmental attention: mercury, lead, and cadmium, because they cause very serious environmental problems. These three elements and their environmental problems are presented in the next three sections. They are illustrative and typical for the heavy metal pollution problems.

As already mentioned, there are 18 elements that require particular attention in environmental management due to their ecotoxicological effect (see also Jørgensen and Fath, 2011). They are (13) aluminum, (24) chromium, (28) nickel, (29) copper, (30) zinc, (33) arsenic, (34) selenium, (38) strontium, (47) silver, (48) cadmium, (50) tin, (51) antimony, (52) tellurium, (56) barium, (80) mercury, (81) thallium, (82) lead, and (83) bismuth. A short overview of the environmental and ecotoxicological problems of these 18 elements is given below, including their applications.

Table 6.2 provides an overview of the concentration in the earth's crust, in average soil, and in average seawater. If an environmental concentration significantly higher than these concentrations in the Earth's crust, in average soil, and in seawater is recorded, then there is a high probability for an ecotoxicological effect. The last column of the table gives the most important applications of the element and its compounds.

Table 6.3 gives the LD_{50} and LC_{50} values as expressions for the toxicity (see Section 3.2). The third column in this table indicates whether there

TABLE 6.2

Typical Concentrations and the Major Applications of the 20 Elements That Have Particular Environmental Concerns

Element	Concentration in Earth's Crust (mg/kg)	Concentration in Soil (mg/kg)	Concentration in Seawater (µg/L)	Applications
13. Al	81,000	10,000–300,000	1	Construction
51. Sb	0.5	1	0.3	Alloys, plastic
33. As	5	1–40	2.6	Chemicals
56. Ba	425	500	30	Chemicals, glass
82. Pb	16	20	0.02	Batteries, soldering, alloys
48. Cd	0.55	0.06	0.1	Surface treatment, dyestuff
29. Cu	100	20	0.9	Cables, wires, construction
24. Cr	300	100	0.05	Surface treatment, alloys
80. Hg	0.065	0.08	0.15	Instruments, dental use
28. Ni	100	40	6.6	Alloys, surface treatment
34. Se	0.09	0.2	0.05	Glass, instruments, dyestuff
47. Ag	0.1	0.1	0.3	Photochemicals, electronic components, ornaments
38. Sr	250	315	8100	Chemicals
81. Tl	1	0.1	0.01	Electronic, alloys
52. Te	0.002	0.001	0.001	Alloys
50. Sn	40	10	3	Alloys, soldering, chemicals
83. Bi	0.02	1	0.03	Alloys, chemicals
30. Zn	40	50	10	Surface treatment, alloys chemicals

are environmental effects and risks that require particular attention. For instance, it is indicated that unacceptably high concentrations of arsenic can be found at some locations, particularly in groundwater (see Murphy and Guo, 2003). Owing to the toxicity included in the carcinogenic effect of arsenic, treatment of the water is absolutely required. An effective treatment of water is possible by precipitation/coagulation and ion exchange.

Table 6.3 is a first course overview. The toxicology and ecotoxicology are of course much more complex; it would not be possible to summarize them in one table. It is therefore strongly recommended that much more information be sought, if there is the slightest suspicion of an ecotoxicological effect of actual concentrations of the 18 elements listed in Tables 6.3 and 6.4.

TABLE 6.3

Overview of Toxicity Expressed by LD_{50} (Lethal Doses Giving 50% Mortality) and LC_{50} (Lethal Concentration in Medium, Water, or Air, Giving a Mortality of 50%)

Element	LD_{50}	LC_{50}	Ecotoxicological Attention
13. Al	770 (mice)	3900 (zooplankton)	Low ecotoxicological effect
51. Sb	100 (rats)	9000 (fish)	Industrial exposure
33. As	41 (rats)	74,000 (cyclops)	Groundwater, carcinogenic effect nerve inflammation
56. Ba	8–23 (rats)	14,500 (cyclops)	Heart problems, diarrhea
82. Pb	130 (rats)	6 ppm (air, rats)	Anemia, brain damage
48. Cd	80 (rats)	65 (cyclops)	Itai-itai, kidney damage
29. Cu	220 (rats)	10 (cyclops)	Highly toxic to plants
24. Cr	3250 (rats)	100,000 (shrimps)	Carcinogenic effect as Cr(IV)
80. Hg	8 (mice)	5 (cyclops)	Central nerve system, mental retardation, teratogenic effect
28. Ni	1620 (rats)	510 (cyclops)	Allergy, carcinogenic and teratogenic effects
34. Se	7 (rats)	2500 (cyclops)	Liver and kidney damage
47. Ag	129 (mice)	30 (zooplankton)	Skin damage
38. Sr	148 (rats)	125,000 (cyclops)	–
81. Tl	16 (rats)	–	–
52. Te	83 (rats)	–	Teratogenic effect
50. Sn	41 (mice)	55,000 (zooplankton)	Liver and kidney damage
83. Bi	13 (rats)	–	Liver and kidney damage
30. Zn	975 (rats)	10^2 (zooplankton), 10^4 (fish)	Minor carcinogenic effect

TABLE 6.4

Lead in Food

Food Items	Typical Lead Concentration (mg/kg Fresh Weight)		
	England	Holland	Denmark
Milk	0.03	0.02	0.005
Cheese	0.10	0.12	0.05
Meat	0.05	<0.10	<0.10
Fish	0.27	0.18	0.10
Egg	0.11	0.12	0.06
Butter	0.06	0.02	0.02
Oil	0.10	–	–
Corn	0.16	0.045	0.05
Potato	0.03	0.1	0.05
Vegetable	0.24	0.065	0.15
Fruit	0.12	0.085	0.05
Sugar	–	0.01	0.01
Soft drink	0.12	0.13	–

Source: Jørgensen, S.E., *Principles of Pollution Abatement*, Elsevier, Amsterdam, 520pp., 2000.

The next three sections discuss the properties and the effects of the three most environmentally important elements, which are repeatedly emphasized when we discuss environmental problems of heavy metals: mercury, lead, and cadmium. The latter is not a heavy metal as its specific density is less than that of iron, but cadmium is anyhow often considered in environmental context as a heavy metal.

6.2 Mercury

Mercury is an extremely toxic element. It has no role as a biological element beyond its toxicity and is not essential for any organism. It is element number 80 in the periodic table and has an atomic weight of 200.59. Mercury has caused one of the most significant environmental catastrophes in history. It was discharged with wastewater from a chemical factory in the Minamata Bay, Japan, in the 1950s (see Newman and Unger, 2003). The discharged mercury accumulated in the sediment where it reacted microbiologically with organic matter and formed methyl mercury ions and dimethyl mercury, which has a low boiling point and can be transferred from the sediment to the water phase or even to the atmosphere.

Organic mercury compounds can be taken up by fish, where a 3000 times higher concentration than in water has been recorded. These processes explain why the mercury concentration was very high in the fish caught in the Minamata Bay. As a result of mercury poisoning of the population living in the surroundings of the bay, hundreds of people died and thousands were invalidated. In the 1950s, the victims were considered to have a new unknown disease, the Minamata disease. Later in the 1960s, it was detected that the disease was caused by a very high mercury concentration, particularly in the fishermen's families. The accumulation in the sediment, the biomagnifications through the food chain (water–phytoplankton–zooplankton–fish–fisherman), and the ability of mercury to form organic compounds by microbiological reactions explain the emergence of the Minamata disease and mercury poisoning caused by the high concentration of mercury in the fish caught in the Minamata Bay and its vicinity.

The use of mercury has declined in the last few decades due to strict environmental legislation rooted in the extreme toxicity of mercury. Mercury compounds were previously applied as fungicides, as dyestuff (Cinnabar red), and for the production of chlorine, but these applications are all now banned in all industrialized countries; however, chlorine and sodium hydroxide are unfortunately still produced in some developing countries through the use of the method based on a dripping mercury electrode, which implies that mercury is discharged with wastewater into the environment. Mercury is still applied by dentists, but much less than previously, and for some electrical instruments. The major mercury contamination today is, however, caused by coal-fired power plants due to the presence of a small concentration of mercury in all coals. A total of 3300 tons of mercury per year is discharged as air pollution into the atmosphere from coal-fired power plants and incineration of solid waste according to Nescaum (2003). The natural emission of mercury is about 1500 t/year, mainly from volcanic activity.

A significant mercury pollution also originates from small-scale gold-refining facilities, which can be found in 55 countries. The total emission from the small-scale refineries is hardly known, as they are not under proper control. It is however estimated that the total mercury emission from these facilities may be as much as 2000 t/year. The burning of mercury–gold amalgam in these "gold shops" can have serious health effects both locally and globally and it is therefore of utmost importance to take this mercury pollution source more seriously. It is possible to install a cost-moderate device to capture the mercury aerosol particles emitted from the gold shops. The use of this equipment, which is denoted MCS—mercury capture system—has, however, been relatively modest up to now.

Cement plants are the fourth largest emitter of airborne mercury in the United States. It is estimated that mercury contamination from cement plants alone in the United States is about 20 tons. Facilities that recycle auto scrap are another big source of mercury pollution, historically pouring 10–12 tons of mercury into the air every year in the United States.

The most important acute effects of mercury are on the lungs and the central nervous system. The chronic effect is complete damage of the central nervous system, which was also the main symptom of the fishermen in Minamata. In addition to the damage of the central nervous system, the victims also lose weight and appetite. The LD_{50} value for mice is 5 mg/kg and WHO recommends a maximum concentration of 1 ppm in food items and fish is mainly the focus. Mercury has furthermore teratogenic and genetic effects.

The main contamination route of mercury from the environment to human passes through the aquatic food chain, where bioaccumulation from water to fish, including the processes via the sediment and biomagnifications through the entire aquatic food chain, takes place. Terrestrial contamination by mercury that takes place, for instance, close to small-scale gold refineries or close to coal-fired power plants is furthermore frequently observed. Similar processes as in the aquatic ecosystems resulting in bioaccumulation and biomagnifications take place in this case. For details about all the effects, see Jørgensen et al. (1991, 1998, 2000).

The evaluation of mercury pollution requires the environmental management to consider the environmental processes in which mercury can participate. The formation of organic mercury compounds mainly in the form of methyl and dimethyl mercury is in this context extremely important and these processes should therefore be included in environmental considerations. It implies that an ecotoxicological model of the distribution and effect of mercury must encompass these processes.

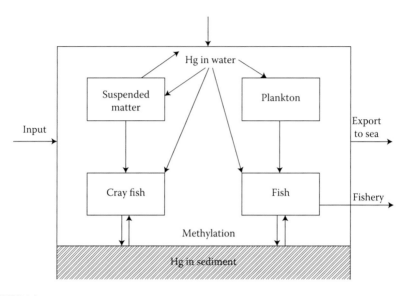

FIGURE 6.1

Conceptual diagram of the mercury model for Mex Bay, Egypt. Note that the methylation is considered for the release of mercury from the sediment. Also note that the mercury in fish is determined by uptake directly from the water (including the concentration of organic mercury) and through the food chain—here only indicated as water–phytoplankton–fish but 1–2 additional steps could be considered.

Figure 6.1 shows a conceptual diagram of mercury distribution and effect in Mex Bay, close to Alexandria. The details of the model can be found in Jørgensen and Bendoricchio (2001). From the figure, it is clear that the methylation processes are included and that the uptake from water and bio-magnifications through the food chain are included to determine the most important state variable of the model: mercury concentration in the top-carnivorous fish (tuna fish), which is an important fish consumed in the Mediterranean region. Figure 6.2 shows the input/output processes that are included in this model.

The uptake of mercury by fish is highly dependent on the pH due to the pH dependence of the solubility of mercury. This dependence is of course mainly of interest when the mercury pollution is inorganic. Figure 6.3 shows the dependence on pH for the uptake of mercury by fish when exposed to 1.5 ppm inorganic mercury (see Jørgensen, 2000). The uptake of mercury from the food is very different for organic mercury and inorganic mercury, namely, 90% and 20%, respectively. This difference entails that a mercury contamination model should follow and include the processes for both inorganic and organic mercury to account for the final concentration in top-carnivorous fish as a result of mercury contamination, whether the discharge is inorganic or organic mercury. The transfer between the two forms should be included, too.

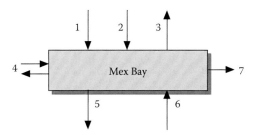

FIGURE 6.2
A biogeochemical model of the distribution of mercury in an aquatic ecosystem, which must include all inputs and outputs. The numbers indicate (1) discharge from waste and tributary to the bay, (2) deposition of mercury from the atmosphere (for instance, from coal-fired power plants), (3) evaporation of mercury, (4) input and output from the open sea, (5) sedimentation, (6) release from the sediment, and (7) fishery.

Mercury pollution caused by coal-fired power plants is reduced for the part of mercury adsorbed to particulate matter by methods used for the treatment of air pollutants to the extent that they are removing particulate matter (see Chapter 11). However, oil and natural gases will be depleted within 60–80 years, which implies that the emission of mercury from coal-fired power

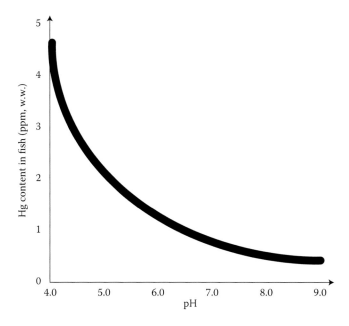

FIGURE 6.3
The concentration of mercury in fish expressed as ppm wet weight versus pH. The concentrations are measured at steady state and for a concentration of mercury chloride in the water of 1.5 ppm (mg/L). The values are based on many measurements for a wide spectrum of different fish species and the concentration in the fish is therefore indicated as a range.

plant will continue if the power plants continue to use coal as fossil fuel. A significant part of the emitted mercury is not adsorbed to the particulate matter. The question is whether, due to the emission of carbon dioxide, we can continue to use coal-fired power plants after 2050–2080. Today, the emission of mercury is indicated as above 3300 t/year, and it is therefore a question whether we can use coal-fired power plant after the oil and natural gas have been depleted, not only due to the continuous emission of carbon dioxide but also due to the continuous emission of mercury, which is an inevitable consequence of the use of coal. This problem has not yet been examined in detail to give a clear answer—so, this question is still open and requires further attention. Probably it will become necessary, due to the greenhouse effect of carbon dioxide, to shift partially or completely to alternative energy sources before 2080. Furthermore, coal has the highest emission of carbon dioxide per kilojoule and will therefore be the fossil fuel that will and should be phased out first.

The mercury pollution causing wastewater problems can be solved by precipitation and ion exchange (ion exchangers with a particular high efficiency of mercury removal are available). Furthermore, adsorption on activated carbon removes inorganic and organic mercury compounds more effectively. The above-mentioned MCS uses the high adsorption of mercury compounds on activated carbon. In this context, it should also be mentioned that bioremediation methods are also able to reduce mercury concentration in contaminated soil.

The solution of mercury pollution for small-scale mercury refineries, MCS, has been already mentioned above. It is important that the legislation for these small-scale gold refineries be tightened and that the use of MCS is made compulsory.

In conclusion, mercury pollution problems can be solved by today's technology except for the emission of mercury from coal-fired power plants, which requires a particular attention if the use of coal as energy source continues to grow, unless the problem is solved by shifting to alternative energy sources. It is furthermore expected that the general restrictions on the use of mercury compounds will be tightened in the years to come.

6.3 Lead

Lead is found in all environmental components—both living and nonliving. The metal lead has dispersed worldwide due to its long-term use in gasoline, batteries, solders, pigments, ammunition, paints, ceramic, and even piping. It is found for instance on glacial ice and snow in Greenland, which is one of the most uncontaminated places on Earth. Many toxic

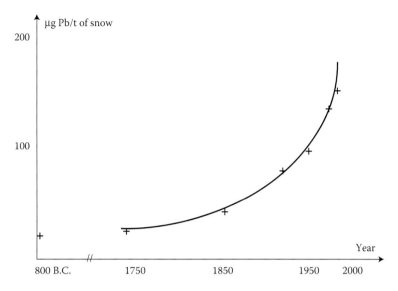

FIGURE 6.4
Lead concentration in glacial ice, Greenland, as function of time, from 800 B.C. to 2000 A.D. The lead concentration has increased 10 times during the last 250 years mainly due to combustion of leaded gasoline. (After Jørgensen, S.E., *Principles of Pollution Abatement*, Elsevier, Amsterdam, 520pp., 2000.)

substances are generally widely dispersed and a global increase in the concentration of heavy metals and pesticides has been recorded, as exemplified for lead in Figure 6.4. The concentration of lead is shown in μg/t as a function of time, which can be found by the analysis of ice cores. As it can be seen, the lead concentration has since the mid-eighteenth century increased 10 times—from about 20 μg/t to about 200 μg/t of snow. The dispersion of lead is caused by the many uses of this metal: in mining and smelting, in batteries, in lead-based paints, in electronic devices, in leaded gasoline, and in shots used in hunting and target shooting. The last two applications were phased out in most industrialized countries more than 25 years ago, but lead gasoline is still in use in many developing countries. In the United States, the dominant use of lead today is in the manufacture of batteries (see Laws, 1993). The global dispersion of lead is particularly caused by the combustion of leaded gasoline, which today is inconceivable because the organic lead chemicals were applied in the gasoline to obtain a sufficiently high octane number at the lowest cost, and the less harmful alternatives could only have increased the gasoline price about a quarter of an American cent per liter.

The concentrations of lead in food items are shown in Table 6.4 to illustrate the presence of lead in our food—another illustration of the consequences of the global dispersion of lead. The concentrations in the table are taken from the mid-1980s, that is, before the introduction of lead-free gasoline had shown any significant effect. The differences between the three countries are explained by the differences in the traffic (and population) densities.

The concentration of lead in completely uncontaminated water is about 1 ng/L while concentrations of 20 ng/L are often found, when only minor discharge of lead has taken place. In contaminated and very contaminated water, a lead concentration of 100–200 ng/L is often found.

Heavy metals are dispersed globally, but the regional concentrations of most heavy metals may of course be much higher than the average global concentration. The relationship between a global and a regional pollution problem and the role of dilution for this relationship are illustrated in Table 6.5 where the ratios of heavy metal concentrations in the river Rhine and in the North Sea are shown. Note that the amount of nickel and lead used in the region of the river Rhine is the same but the ratio is 70 times higher for lead than for nickel due to the application of leaded gasoline, which disperses the lead uncontrollably, while nickel has more closed applications, which also allow recycling of the metal. Also note that lead is transported in the atmosphere primarily in the particulate phase according to Haygarth and Jones (1992).

The toxicity of Pb is mainly associated with the free ions, Pb^{2+} (the +2 oxidation state), but lead can form complexes with hydroxide ions, carbonate, chloride ions, and many organic compounds, for instance, humic acid and amino acids. The toxicity of the complexes is generally lower than the toxicity of the free ions because the uptake of the complexes by organisms is slower than the uptake of the free ions. It implies that in every case study it is necessary to determine the concentrations of lead as free ions by analytical methods or by chemical calculation (see Chapter 7) and to determine the toxicity in the form of various complexes. In general, the toxicity of lead declines with higher concentration of the hardness ions, namely, calcium and magnesium. Lead shows bioaccumulation and biomagnification as other heavy metals, which is more pronounced at lower pH because the

TABLE 6.5

Heavy Metal Pollution in the River Rhine (Since 1985)

	Discharges in the River Rhine (t/year)	Ratio of Concentration in the River Rhine to Concentration in the North Sea
Cr	1000	20
Ni	2000	10
Zn	20,000	40
Cu	200	40
Hg	100	20
Pb	2000	700

Source: Jørgensen, S.E., *Principles of Pollution Abatement*, Elsevier, Amsterdam, 520pp., 2000.

solubility and the relative concentration of the free ions are increasing with decreasing pH.

Contaminated aquatic ecosystems have a significant elevated concentration of lead in the sediment. Generally, the sediment has higher concentrations of heavy metals than water, and as it is possible to analyze sediment core (see Forstner and Wittmann, 1979; Jørgensen and Fath, 2010), it is possible to find the contamination of heavy metals as functions of time, provided the settling rate is known or can be estimated. This is particularly informative in the case of lead, because the use of leaded gasoline started shortly after the Second World War and was banned in the industrialized countries before or around the mid-1980s. It entails that the sediment, from about 40 years when leaded gasoline was used, will show a particularly high lead concentration, which of course facilitates the dating of the sediment. The lead concentration in sediment is usually 10–200 mg/kg dry weight, but as much as 3000–10,000 mg/kg dry weight can be found in contaminated areas.

A filter feeding bivalve mollusk shows a contamination of lead (and other heavy metals) that is proportional to the concentration in the sediment. The proportional constant is dependent on the composition of the sediment, but it is frequently between 0.01 and 0.05—the highest values for sediment with a high concentration of organic matter (see Jørgensen and Fath, 2011).

A major source of lead exposure and toxicity for wild birds is the ingestion of lead-based ammunition. For birds, concentration of lead is in the order of 0.2 mg/kg dry weight, while a toxic effect would correspond to 100 times as much and death to 250 times as much. LD_{50} for rats is 130 mg/kg dry weight (see Chapter 3). There is a primary lead shot poisoning from direct ingestion of lead-based ammunition and a second lead shot poisoning when birds (and of course other animals) ingest lead shotgun pellets and bullet fragments embedded in the flesh of dead or wounded animals shot with lead-based ammunition.

Lead is bound to SH–groups in the proteins and can therefore generally be taken up more effectively by plants with high concentrations of proteins than by plants with low protein concentrations (see Chapter 12 about bioremediation). Removal of lead from areas that have been applied for target shooting with ammunition containing lead is possible by bioremediation (see Chapter 12 and Mitsch and Jørgensen, 2004).

Similar to other heavy metals, the uptake of lead from the food is relatively low—about 7%–10% (see Jørgensen and Bendoricchio, 2001; Newman and Unger, 2003), while lead is taken up from the atmosphere by the lungs with a higher efficiency. Figure 6.5 shows a steady-state model for the uptake for an average European in the mid-1980s, just before the use of leaded gasoline was banned for all new cars. The uptake from food is as seen 10%, namely, 30 μg per day out of 300 μg per day that is in the food (compare with Table 6.4). The amount of lead in 20 m^3 of air that was used daily for respiration was about 25 μg, but as it was taken up with an efficiency of 50%, as much as 12.5 μg lead per day accumulated in the body of an average European 25 years ago

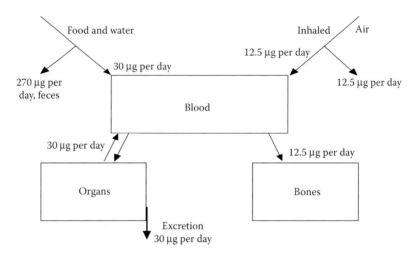

FIGURE 6.5
A steady-state model of an average European in 1985. The uptake from food was 30 µg per day out of 300 µg per day (10% efficiency) and 12.5 µg lead per day by respiration (uptake efficiency 50%). The excretion balances the uptake from the food and the lead taken up by respiration is accumulated in the bones.

due to direct atmospheric pollution. 30 µg lead per day is excreted mainly through urine and therefore 12.5 µg lead per day is accumulated in the body, mainly in the bones, where their effect fortunately is very minor. Owing to the reduced use of leaded gasoline, the average European will today have less lead in the body. Food contains roughly half as much lead today than 25 years ago and the atmospheric pollution is also half the level today compared with that in 1985. It means that the amount of lead from food today is 15 µg lead per day and from respiration 6.25 µg lead per day. The excretion is also reduced to a level about 15–20 µg lead per day, which implies that less lead is accumulated in the bones today. The indicated amounts of lead for an average European today are found by the use of a model that was calibrated and validated by the use of the amounts from 1985—it means the values that are shown in Figure 6.5.

The spectrum of toxicological and ecotoxicological effects of lead is very wide. Acute toxicity of lead causes headache, irritability, and loss of appetite. Chronic toxicity leads to brain damage (memory is reduced), anemia, damage of liver and kidney, and possibilities for cancer tumors in the kidney. It has been shown that if children are exposed to high lead concentrations, it will have a pronounced effect on their learning ability according to Bellinger et al. (1986). Elevated lead concentrations also have teratogenic effects (see Om Metaller, 1980) and prenatal lead exposure has been demonstrated to be associated with increased risk for malformations according to Needleman et al. (1984). Genetic effects on animals have also been observed (see Om

Metaller, 1980). Epidemiological studies have shown that lead is related to the risk for elevated blood pressure according to Schwartz et al. (1986).

Significantly different lead concentrations have been found in the blood of people living far from towns and cities and those living in urban areas. Indians in the Amazon have about 8 ng lead per gram of blood. Farmers living on the countryside in Europe have about 50 ng/g of blood and people living in very big cities (New York, for instance) had about 150 ng/g of blood (see Jørgensen et al., 1991). The highest concentrations of lead have been found in policemen who regulated traffic in industrialized countries before lead was phased out: 150–700 ng/g blood. Lead in clean air is as low as 0.0005 ng/L, whereas in urban areas it is mostly between 2 and 25 ng/L (see Francis, 1994). In much polluted urban areas, the concentration may even reach 50 ng/L, which could give a daily uptake of 0.345 mg per person and a lead concentration in blood about 720 ng/g blood; see Francis, 1994). The model in Figure 6.5 could be used to find these values in the blood, when the air pollution level is known.

Lead in soil shows a similar wide range of values. Generally, a lead concentration of 0.2–5 mg/kg dry matter is found in agricultural areas according to Prost (1995), but a high lead concentration of 100 mg/kg dry matter is found in particularly contaminated soil (see Hutchinson et al., 1994), and in some extreme cases, concentrations as high as 10,000 mg/kg dry matter have been found (see Kabata-Pendias, 1986). A typical soil in an industrialized country contains about 20 mg/kg dry matter of lead and 95% of soil samples randomly sampled will show concentrations between 10 and 60 mg/kg dry matter. Lead concentrations above 1000 mg/kg dry weight in soil can kill earthworms and springtails.

The half-life of lead in humans is about 6 years (whole body) and about three times as much for the lead in skeletal tissue. It has been shown that skeletal burdens of lead increases linearly with age, while the nonskeletal burden is eliminated faster (as indicated, the whole-body lead has a half-life of 6 years, which means that half of the nonskeletal lead is exchanged faster than once every 6 years). It is possible to reduce the body burden for patients clinically affected by lead by EDTA (see Nigel, 1994).

Two major sources of lead pollution, namely, combustion of leaded gasoline and the use of lead ammunition have been eliminated in most industrialized countries by environmental legislation. The use of lead in ceramic and paints has similarly been reduced significantly by legislation or by agreement between environmental agencies and the industry. Lead pollution has therefore been reduced considerably during the last 25–30 years due to environmental legislation and also due to the treatment of wastewater and contaminated smoke and air. Environmental technological solutions have been increasingly applied for the treatment of water and air. A description of the environmental technological methods for the removal of heavy metals is provided in Chapter 11. The most important methods are precipitation through the use of calcium hydroxide and application of ion exchange for treatment of wastewater.

Contaminated land has also been treated by either environmental technological methods or bioremediation (see Chapter 12). A few cases of successful application of cleaner technology (replacing lead by less harmful components) have also been reported in the journal *Cleaner Technology*. The results of these efforts are encouraging for the use of a consequent, integrated environmental management. When environmental legislation, environmental technology, cleaner technology, and ecotechnology work hand in hand, it is possible to achieve good pollution abatement results, as the lead pollution problem has demonstrated.

Like all pollution problems, lead contamination is complex. It is briefly discussed in this section as to how to go around this complexity to be able to propose an integrated and holistic environmental management and thereby solve the problems properly. The discussion is in principle valid for all pollution problems; compare with Chapter 1. The following crucial questions need to be answered in the case of lead contamination:

1. Which are the forms of lead that have lead-free ions, or which complexes in which concentrations?
2. What are the sources of the problem? Quantitatively?
3. Which is the diagnosis for lead contamination, meaning the relationship between the quantities of lead (the impact) and the effects on all hierarchical levels (from landscape to cells)?
4. Which possibilities do we have to eliminate which sources? Will that be sufficient to solve the problem?
5. How can we best combine the methods to solve the problem?
6. To what extent have we solved the problem (the diagnosis can be used here)?

It is recommended to consider the following points to be able to arrive at an answer to the crucial questions:

1. The forms of lead can either be found analytically or by chemical calculations (see Chapter 7, where straightforward chemical calculations are clearly shown for heavy metals with illustrative examples).
2. It is advantageous to set up a mass balance. Figure 6.6 shows a mass balance for lead contamination of 1 ha Danish agricultural land. The mass balance clearly reveals the important sources. Atmospheric fallout is the dominant source, although it may also be beneficial to reduce lead contamination coming from sludge and fertilizers. Particularly, the first of these two could be eliminated.
3. It is possible to eliminate air pollution of lead as it has been discussed by phasing out the use of lead in gasoline. The second most important

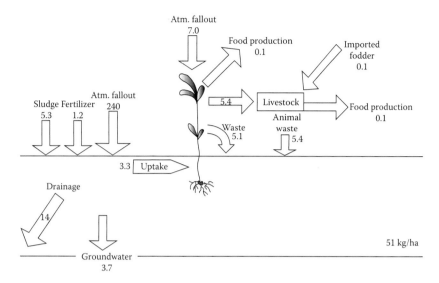

FIGURE 6.6
Lead balance of average Danish agriculture land. All rates are in g Pb/ha year.

source of air pollution in many countries is coal-fired power plants, which can be eliminated by shifting to other forms of fossil fuel or to alternative energy sources. This change of the energy policy will, however, be beneficial on a long-term basis, but it will require investment on a short-term basis and it is therefore a political question. The phasing out of lead in gasoline may be sufficient. The core question is: How much will the lead contamination in food be reduced if we reduce the atmospheric fallout by so much? The answer requires calculations and sometimes the use of models. Lead models have been developed by Jørgensen (see Jørgensen and Bendoricchio, 2001; Jørgensen and Fath, 2011). Lead models considering hydrodynamics, bioconcentration, bioaccumulation, excretion, and sedimentation can furthermore be found in Lam and Simons (1976) and Aoyama et al. (1978). But another core question is whether we are able to reduce the concentration in food or in sediment or in soil to desired levels, and if so, would this reduction be sufficient to reduce or even eliminate the effect? This question will require a comprehensive overview of the toxicological and ecotoxicological literature about lead and its effect. If the knowledge about the effect is not available, it is necessary to perform bioassay (see Newman and Jagoe, 1996; Landis and Yu, 1995). These references describe in detail how to perform a bioassay for heavy metals, including lead.

4. The answer to the fourth question is rather complex, but it could be mentioned that it is necessary after the extensive use of models

and indicators to combine solutions for all the sources of pollution, which includes environmental technology, ecological engineering, environmental legislation, and cleaner technology. The first toolbox is used for industrial pollution, the second toolbox for the restoration of contaminated land, the third toolbox to ban the use of lead in gasoline and for ammunition, and the fourth toolbox for industrial pollution where lead obviously could be replaced by other metals or even other types of material.

6.4 Cadmium

Cadmium is used for metal surface treatment, as a stabilizing agent in plastic, and in many alloys. Phosphorus fertilizers have a relatively high cadmium concentration—on the order of 10–80 mg/kg. They are, along with coal-fired power plants that emit cadmium, the most important source of global cadmium dispersion. More than 1000 t of cadmium is globally dispersed from coal-fired power plants (see Om Metaller, 1980). Cadmium was previously applied in ceramic, but this use of the very toxic metal is now banned in most industrialized countries.

From 1940 to 1960, Japanese in the Toyama Prefecture were poisoned by cadmium in their rice, because the river water used for irrigation was contaminated by cadmium from a cadmium mine (Newman and Unger, 2003). Cadmium replaces calcium in the bones, which makes them soft. It causes extreme pain and therefore the disease was named itai-itai, which means ouch-ouch. When it was discovered that cadmium was the cause of the disease, the mine waste was controlled; however, by then, several thousand of farmers had already suffered from the very painful disease.

Cadmium, like other heavy metals, shows biomagnification and bioaccumulation. It is also like most other heavy metals accumulated in the sediment of aquatic ecosystems. It entails that the dispersion of cadmium in aquatic ecosystems can in most cases be best determined by analyses of the sediment where relatively higher concentrations are found (see Jørgensen, 2000). Similar to lead (see Section 6.3), it is possible to determine the history of cadmium emission by analyses of sediment cores. The phosphorus fertilizer industry had discharged cadmium-containing waste to The Little Belt in Denmark for several decades. An environmental impact assessment was carried out for The Little Belt by analyses of a large number of sediment cores, whereby the history of the contamination was determined. Fortunately, cadmium can form complexes with chloride, which are less toxic than cadmium ions (see Jørgensen et al., 1991, 1997). The complexes are at the same time more soluble (see Chapter 7), which means that less cadmium is transferred to the sediment and more to the open sea where it is diluted significantly. The discharge of cadmium waste was stopped as a result of the investigation, but it was also concluded that the contamination of The Little Belt was less than expected probably due to the formation of cadmium–chloride complexes.

Owing to the high cadmium concentration in phosphorus fertilizers, there is a risk for cadmium contamination of agricultural land, particularly due to intensive agriculture. Cadmium is taken up by plants (see bioremediation in Chapter 12). A cadmium model has been developed to be able to relate the cadmium concentration in crops as a function of cadmium contamination by the use of fertilizers (see Jørgensen and Fath, 2011).

It is important to make a regional mass balance to identify the sources and dispersion of all toxic substances and particularly for the most toxic heavy metals so as to assess whether a specified toxic substance would reach an unacceptable high concentration and thereby do most harm. Jørgensen and Fath (2011) have exemplified regional mass balances. Figure 6.7 shows one example, namely, a cadmium balance for Danish agricultural land, but similar balance can also be found for lead (see Section 6.3), mercury, and copper, for other regions. The cadmium pollution of agricultural land is due to the use of fertilizers (1.7 g/ha year), from the use of wastewater sludge (0.18 g/ha year), and from air pollution, dry deposition, and rainwater (1.7 g/ha year). 1.7 g/ha year is taken up from the soil by plants and 0.2 g/ha year

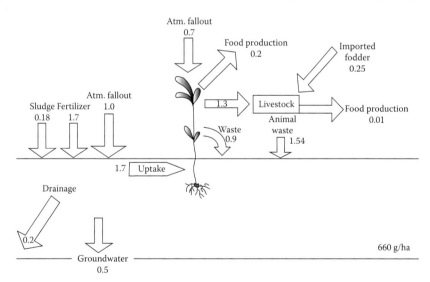

FIGURE 6.7
Cadmium balance for Danish agricultural land. The accumulated amount is about 660 g/ha, while the annual input by fertilizers is 1.7 g/ha and by air pollution another 1.7 g/ha, of which the 1 g is accumulated directly on the bare soil and the 0.7 g on the plants. The amount from fertilizers corresponds to the amount taken up by plants. 1.3 g/ha of the cadmium contamination of plants and 0.25 g/ha of cadmium in imported fodder give and accumulation of 1.55 g cadmium/ha year in the domestic animals but only the 0.01 g/ha will end up as cadmium contamination of the animal products.

contaminates vegetable products, while only 0.01 g cadmium/year contaminates animal food even if the cadmium in imported fodder is considered. Most of the cadmium entering domestic animals go back to the agricultural land through animal waste. Totally, about 3.8 g cadmium is added to the agricultural land per year but 0.7 g cadmium/ha year is transported by drainage water and groundwater to other sites (ecosystems) and 0.2 g/ha year will be removed during the harvest of vegetables. The net accumulation in agricultural soil is therefore 3.1 g/ha year. For details about this illustrative example, see EPA, Denmark (1979), Hansen and Tjell (1981), and Jensen and Tjell (1981) and compare with Chubin and Street (1981). Mass balance can be used to calculate the effect of fertilizers containing less cadmium and the result of a reduction of cadmium discharged to the atmosphere. If more complex pollution abatement strategies are applied, it is necessary to apply an ecotoxicological model with the state variables, processes, and transfers as shown in Figure 6.7 as the core of the model.

Cadmium has carcinogenic and teratogenic effects. It is highly toxic as indicated by the LD_{50} value for rats—70–90 mg/kg. LD_{50} for cadmium in smoke (as cadmium oxide) by inhalation is 500 min mg/m^3. The cadmium concentration in rice that caused the itai-itai disease in Japan was about 1 mg/kg.

The uptake of cadmium from food is about the same as that for other heavy metals—7%–10%. Cadmium is accumulated mainly in the kidney. The total cadmium concentration in the body increases with age, as the uptake is higher than the excretion mainly through urine. Excretion follows a first-order reaction with a coefficient of 0.0001 1/24 h. At age 50, the cadmium content in the body will be about 25 mg, of which 8–9 mg are accumulated in the kidney and 3 mg in the liver (see Om Metaller, 1980). The daily accumulation is about 2–3 μg.

It is possible to express the cadmium accumulation in the body as a function of time by the following differential equation:

$$\frac{dCd}{dt} = \text{daily uptake} - \text{excretion coefficient} * Cd \quad\quad (6.1)$$

where Cd is the amount of cadmium in the body and the daily uptake is found to be about 10% of the cadmium content in the food and 50% of the cadmium in 20 m^3 of air used for respiration per 24 h. The intake by food is about 25 μg/24 h and the intake by respiration is very minor. Therefore, the equation can be written as

$$\frac{dCd}{dt} = \frac{0.025 * 10}{100} - \frac{0.0001 * Cd \, mg}{24 \, h} \qu\quad (6.2)$$

At steady state, the cadmium concentration becomes 0.0025/0.0001 = 25 mg, which (of course) is close to the average value at the age of 50 years.

6.5 Effects of Organic Compounds

A short overview of the most toxic organic compounds is given below. These compounds are very important from an environmental point of view as they threaten the environment and have several harmful effects. The overview applies the following classification:

A. Petroleum hydrocarbons

B. PCBs and dioxins

C. Pesticides

D. Polycyclic aromatic hydrocarbons (PAHs)

E. Heavy metals

F. Detergents

G. Synthetic polymers and xenobiotics applied in the plastics industry

The properties and the effects of the above-listed seven classes of contaminants that represent the most toxic organic compounds that we are using and the most harmful compounds for the environment are briefly treated. For a more comprehensive coverage of toxic chemical compounds in the environment, the reader can refer to Loganathan and Lam (2012), Newman and Unger (2003), Hoffman et al. (1994), and Schuurmann and Markert (1998).

6.5.1 Petroleum Hydrocarbons

Petroleum hydrocarbons include a variety of organic compounds. Hydrocarbons (compounds composed of carbon and hydrogen) constitute only 50%–90% of petroleum. They are *n*-alkanes, branched alkanes, cycloalkanes, and aromatics. Cycloalkanes usually comprise the largest portion of hydrocarbons in petroleum, while aromatics are usually present to the extent of 20% or less. The characteristic compounds are benzene, alkyl-substituted benzenes, and fused ring polycyclic aromatic hydrocarbons (PAHs; see also this group, D). Some of these aromatics are carcinogenic and these are probably the group that is of greatest environmental concern. In addition to hydrocarbons, petroleum contains sulfur and nitrogen compounds, such as thiophene, ethanethiol, and pyridine derivatives.

Petroleum compounds are emitted or discharged into all spheres. Evaporation removes the lower-molecular weight, more volatile components of the petroleum mixture. Hydrocarbons with vapor pressures equal to that of *n*-octane (0.019 atm at room temperature) or greater will be lost quickly via evaporation.

The lower-molecular-weight hydrocarbons also tend to be the most water soluble, but for the same molecular weight, aromatics are more soluble than cycloalkanes, which are more soluble than branched alkanes with the *n*-alkanes being the most insoluble in water. Petroleum products are discharged directly into the sea by accidents or by violation of international regulations. In a massive discharge of petroleum products, which has been recorded in several ship accidents, most of the petroleum will initially float on the surface of marine waters as a slick. Eventually, most slicks are dispersed widely and form a 0.1-mm-thick layer on the water. If drift results in landfall, coral, or contact with mangrove communities, a disastrous environmental impact may occur.

Fortunately, chemical transformation and degradation processes act on petroleum compounds in the environment. Microbial transformation and photooxidation are of particular importance. The aromatic compounds of petroleum are the most toxic substances in this class of organic pollutants. They are lethal to crustaceans and fish with LC_{50} in the range of 0.1–10 mg/L.

6.5.2 PCBs and Dioxins

PCBs and dioxins are characterized as aromatic compounds with a high content of chlorine. As the names indicate, polychlorinated biphenyls and polychlorinated dibenzo (1,4) dioxins, there are many different individual compounds under these labels. A total of 209 different PCB compounds are known although only about 130 are found in commercial mixtures. Figures 6.8 and 6.9 show the molecular structures of some of the most common PCBs and dioxins.

The applications of PCBs have been quite diverse (capacitor oil, plasticizers, printer's ink, etc.), but due to investigations in the 1960s and the 1970s, in

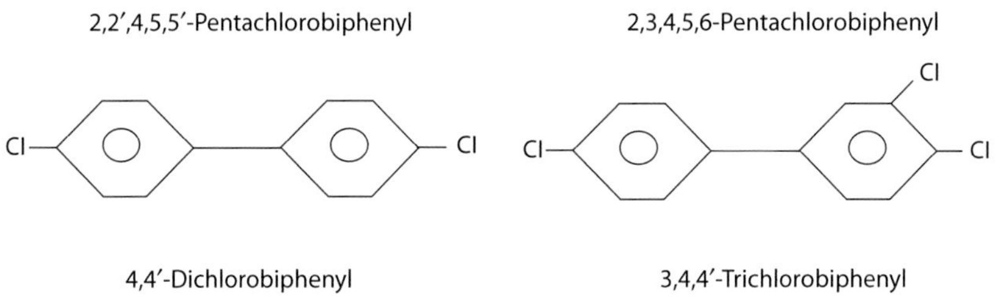

2,2′,4,5,5′-Pentachlorobiphenyl   2,3,4,5,6-Pentachlorobiphenyl

4,4′-Dichlorobiphenyl   3,4,4′-Trichlorobiphenyl

FIGURE 6.8
Molecular structure of four common PCBs.

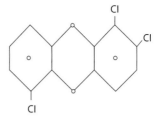

1,2-Dichlorodibenzo (1,4) dioxin   1,2,6-Trichlorodibenzo (1,4) dioxin

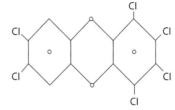

2,3,7,8-Tetrachlorodibenzofuran   1,2,3,4,7,8-Hexachlorodibenzo (1,4) dioxin

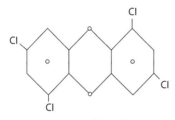

1,3,6,8-Tetrachlorodibenzo (1,4) dioxin

FIGURE 6.9
Molecular structure of five common dioxins.

which it was found that PCBs occurred widely in the environment and significant bioaccumulation took place, voluntary restrictions were introduced and all "open" applications banned.

Dioxins are not deliberately produced but are by-products of chemical processes involving chlorine, for instance, the production of various organochlorine and bleaching of pulp and of combustion processes if chlorine-containing compounds are present.

Both PCBs and dioxins are characterized by low water solubility and high K_{ow} (most components have log $K_{ow} > 5$). Both groups of compounds are very persistent to decomposition processes, which explains why they are strong bioaccumulators, although dioxins have a UV–Vis absorption spectrum that results in significant absorption from solar radiation. Some dioxins have a half-life in the troposphere of a few days.

6.5.3 Pesticides

Pesticides are used to remove, abate, and kill pests, and probably due to their direct use in nature they have been the most criticized environmental contaminants. Usage of DDT and related insecticides accelerated during the 1940s and the subsequent decades until environmental doubt occurred in the mid-1960s. Since 1970, DDT has been banned in most industrialized countries, but it is still used in developing countries, for instance, India, where it has resulted in very high body concentrations in the Indian population. All the chlorinated hydrocarbon insecticides are banned in most industrialized countries due to their persistence and ability to bioaccumulate (K_{ow} is high; K_{ow} = the ratio of the solubility in octanol and the solubility in water).

Pesticides can be divided into the following classes depending on their use and their chemical structure:

Herbicides comprise carbamates, phenoxyacetic acids, triazines, and phenylureas. Insecticides encompass organophosphates, carbamates, organochlorines, pyrethrins, and pyrethroids.

Fungicides are dithiocarbamates, copper, and mercury compounds.

Pesticides are chemically an extremely diverse group of substances, and they only have in common their toxicity to pests. A few of the most important molecules are shown in Figure 6.10. They are mostly produced synthetically, although a natural pesticide pyrethrin has achieved commercial success.

Chlorohydrocarbons are strongly bioconcentrated as already emphasized. In addition, they are very toxic to a wide range of biota, particularly to aquatic biota.

Organophosphates are almost equally toxic to biota, but due to these compounds, lack of persistence, higher solubility in water, and bioaccumulation capacity, they are still in use.

Carbamates are relatively water soluble and have limited persistence. They are however toxic to a wide range of biota. They act by inhibiting cholinesterase.

Pyrethins have a complex chemical structure and high molecular weight. Thus, they are poorly soluble in water and tend to be lipophilic. They are readily degraded by hydrolysis and are more attractive to use than most of the other pesticides due to their very low mammalian toxicity. Phenoxyacetic acid is a very effective herbicide but contains trace amounts of tetrachloro-dibenzo-dioxin.

Pesticides are banned in organic agriculture where they are replaced by other methods, for instance, mechanical and biological methods (use of predator insects).

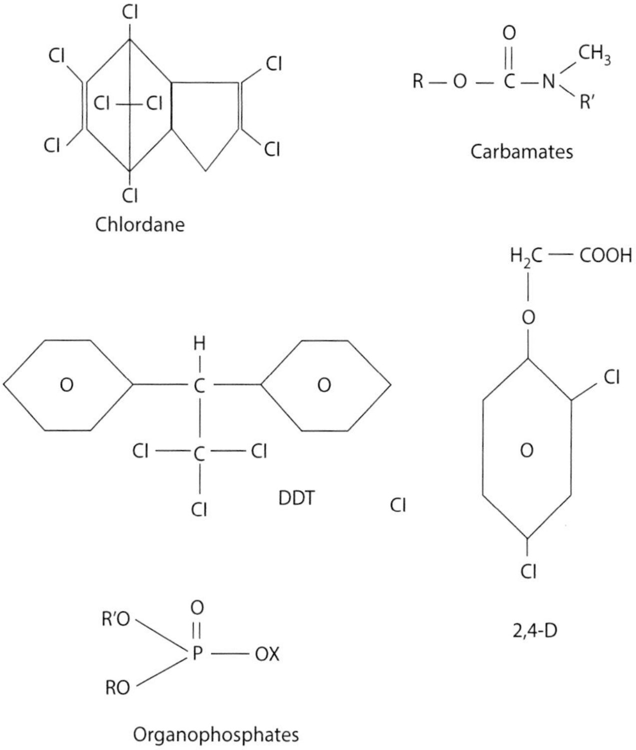

FIGURE 6.10
Molecular structure of five common pesticides.

6.5.4 PAHs

PAHs are molecules containing two or more fused 6C-aromatic rings. They are ubiquitous contaminants of the natural environment, but the growing industrialization has increased the environmental concern about these components. Two common members are naphthalene and benzo(*a*)pyrene (see Figure 6.11). PAHs are usually solids with naphthalene (lowest molecular weight) having a melting point of 81°C.

The natural sources of PAHs in the environment are forest fires and volcanic activity. The anthropogenic sources are coal-fired power plants, incinerators, open burning, and motor vehicle exhaust. As a result of these sources, PAHs occur commonly in air, soil, and biota. They are lipophilic compounds that are able to bioaccumulate.

The low-molecular-weight compounds are moderately persistent; while, for example, benzo(*a*)pyrene with a higher molecular weight persists in aquatic systems for up to about 300 weeks. They are relatively toxic to aquatic organisms and have LC_{50} values for fish in the range of 0.1–10 mg/L. The major

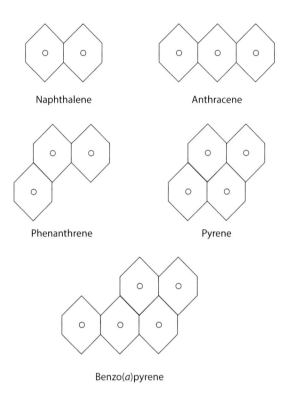

FIGURE 6.11
Molecular structure of five common PAHs.

environmental concern of PAHs is that many PAHs are carcinogenic. It is has been shown (Andersen, 1998) that benzo(*a*)pyrene is an endocrine disrupter, and it cannot be ignored that many more PAHs have adverse environmental effect of disturbing the hormone balance of nature. Human exposure to PAHs occurs through tobacco smoking as well as through compounds in food and the atmosphere.

6.5.5 Organometallic Compounds

Organometallic compounds have metal–carbon bonds, in which the carbon atoms are part of an organic group. The best-known example is probably tetraethyl lead, which is used as an additive to gasoline. It has now been phased out of use in many countries—all industrialized countries—due to its environmental consequences. Organometallic compounds can be formed in nature from metal or metal ions, for example, dimethyl mercury, or are produced for various purposes, as catalysts, for example, organoaluminum; as pesticides, for example, organoarsenic and organotin compounds; as stabilizers in polymers, for example, organotin compounds; and as a gasoline additive, for example, organolead compounds. Organometallic compounds exhibit properties that are different from those of the metal itself and inorganic derivatives of the metal. They have a relatively higher toxicity than the metals in contrast to heavy metal complexes with organic ligands.

Most organometallic compounds are relatively unstable and undergo hydrolysis and photolysis easily. Most organometallic compounds have weakly polar carbon–metal bonds and are often hydrophobic. They therefore only dissolve in water to a small extent and are readily adsorbed onto particulates and sediments.

The most harmful organometallic compounds from an environmental point of view are organomercury, organotin, organolead, and organoarsenic, which are all very toxic to mammals.

6.5.6 Detergents (and Soaps)

Detergents (and soaps) contain surface-active agents (surfactants), which are classified according to the charged nature of the hydrophilic part of the molecule:

Anionic: negatively charged

Cationic: positively charged

Nonionic: neutral, but polar

Amphoteric: a zwitterion containing positive and negative charges

They are produced and consumed in large quantities and are mostly discharged into the sewage system and end up in the wastewater plant. The early surfactants contained highly branched alkyl hydrophobes that were resistant to biodegradation. These surfactants are largely obsolete today having been replaced by linear alkyl benzene sulfonates (LAS) and other biodegradable surfactants.

The toxicity to mammals is generally low for all surfactants, while the toxicity to aquatic organisms is relatively high (LC_{50} from about 0.1 to about 77 mg/L). The toxicity will generally increase with the carbon chain length (see Figure 6.12). Many surfactants bind strongly to soils and sediments, which implies that, to the extent that they are not biodegraded in a biological treatment plant, they will mainly be found in the sludge phase.

6.5.7 Synthetic Polymers and Xenobiotics Applied in the Plastics Industry

Forming a very diverse group of compounds from a chemical viewpoint, synthetic polymers are useful (plumbing, textiles, paint, floor, covering, and as the basic material for a wide spectrum of products) because they are

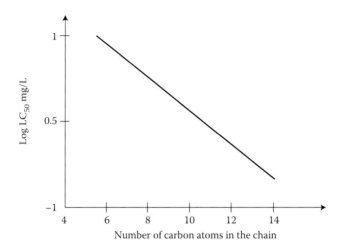

FIGURE 6.12
Log LC_{50} plotted versus number of carbon atoms in the chain for LASs. As seen, increased chain length implies increased toxicity. (From Jørgensen, S.E., *Principles of Pollution Abatement*, Elsevier, Amsterdam, 520pp., 2000.)

resistant to biotic and abiotic processes of transformation and degradation. These properties, however, also cause environmental problems associated with the use of these components. In addition, several xenobiotic compounds are used as additives, softener, stabilizers, and so on in synthetic polymers to improve their properties. Some of these additives are very toxic and may cause other environmental problems; for instance, phthalates are widely used in the plastic industry, and it has been demonstrated that phthalates have effects as endocrine disrupters. After use, synthetic polymers are usually incinerated together with industrial and household garbage (solid waste). The presence of PVC will imply that hydrochloric acid is formed and also dioxins to a certain extent, but this is strongly dependent on the incineration conditions. As it is difficult to separate the different types of plastics, the eventual phase out of the use of PVC has been discussed, but due to PVC's unique properties, this has not yet been decided.

References

Andersen, H.R. 1998. Examination of endocrine disruptors. Master's thesis, DFU, Copenhagen University.
Aoyama, I., Yos. Inoue and Yor. Inoue. 1978. Simulation analysis of the concentration process of trace heavy metals by aquatic organisms from the viewpoint of nutrition ecology. *Water Res.* 12: 837–842.

Bellinger et al. 1986. Low-level lead exposure and infant development in the first year. *Neurobehav. Toxicol. Teratol.* 8: 151–161.

Chubin, R.G. and J.J. Street. 1981. Adsorption of cadmium on soil constituents in the presence of complexing agents. *J. Environ. Qual.* 10: 225–228.

EPA. 1979. *The Lead Contamination in Denmark*. Denmark. 145pp.

Forstner, U. and G.T.W. Wittmann. 1979. *Metal Pollution in the Aquatic Environment*. Springer-Verlag, Heidelberg, Berlin, New York, 490pp.

Francis, B.M. 1994. *Toxic Substances in the Environment*. John Wiley & Sons, New York, 362pp.

Hansen, J.A. and J.C. Tjell. 1981. *The Application of Sludge as Soil Conditioner*, Vol. 2. Polyteknisk Forlag, Copenhagen, pp. 137–181.

Haygarth, P.M. and K.C. Jones. 1992. Atmospheric deposition of metals to agricultural surfaces. In D.C. Adriano (ed.), *Biochemistry of Trace Metals*, Lewis Publishers, Boca Raton, 423pp.

Hoffman et al. 1994. *Handbook of Ecotoxicology*. Lewis Publishers/CRC Press, Boca Raton.

Hutchinson, T.C., C.A. Gordon, and K.M. Meema. 1994. *Global Perspectives on Lead Mercury and Cadmium Cycling in the Environment*. Wiley Eastern Limited, New Delhi, 412pp.

Jensen, K. and J.C. Tjell. 1981. *The Application of Sludge as Soil Conditioner*, Vol. 3. Polyteknisk Forlag, Copenhagen, pp. 121–147.

Jørgensen, L.A., S.E. Jørgensen, and S. Nors Nielsen. 2000. *Ecotox*. CD Elsevier, Amsterdam, Corresponding to 4000 pages.

Jørgensen, S.E. 2000. *Principles of Pollution Abatement*. Elsevier, Amsterdam, 520pp.

Jørgensen, S.E. and G. Bendoricchio. 2001. *Fundamentals of Ecological Modelling*. Third edition. Elsevier, Amsterdam, 530pp.

Jørgensen, S.E. and B. Fath. 2010. *Ecotoxicology*. Elsevier, Amsterdam, 390pp.

Jørgensen, S.E. and B. Fath. 2011. *Fundamentals of Ecological Modelling*. Fourth edition. Elsevier, Amsterdam, 396pp.

Jørgensen, S.E., B. Halling-Sørensen, and H. Mahler. 1998. *Handbook of Estimation Methods in Environmental Chemistry and Ecotoxicology*. CRC Press, Lewis Publishers, New York, 230pp.

Jørgensen, S.E., L.A. Jørgensen, and S. Nors Nielsen. 1991. *Handbook of Ecological and Ecotoxicological Parameters*. Elsevier, Amsterdam, 1380pp.

Jørgensen, S.E., H. Mahler, and B. Halling Sørensen. 1997. *Handbook of Estimation Methods in Environmental Chemistry and Ecotoxicology*. Lewis Publishers, Boca Raton.

Kabata-Pendias, A. 1986. *Effects of Trace Metals Excess in Soils and Plants*. CRC Press, Boca Raton, 330pp.

Lam, D.C.L. and T.J. Simons. 1976. Computer model for toxicant spills in Lake Ontario. In J.O. Nriago (ed.), *Metals Transfer and Ecological Mass Balances, Environmental Biochemistry*, Vol. 2. Ann Arbor Science, Ann Arbor, pp. 537–549.

Landis, W.G. and M.H. Yu. 1995. *Introduction to Environmental Toxicology*. Lewis Publishers, Boca Raton, 330pp.

Laws, E. A. 1993. *Aquatic Pollution: An Introductory Text*. Second edition. John Wiley & Sons, New York, 380pp.

Loganathan, B.G. and P.K.S. Lam. 2012. *Global Contamination Trends of Persistent Organic Chemicals*. CRC Press, Boca Raton, 639pp.

Mitsch, W.J. and S.E. Jørgensen. 2004. *Ecological Engineering and Ecosystem Restoration.* John Wiley, New York, 410pp.

Murphy, T. and J. Guo. 2003. *Aquatic Arsenic Toxicity and Treatment.* Beckhuys Publishers, Leiden, 165pp.

Needleman et al. 1984. The relationship between prenatal exposure to lead and congenital anomalies. *Journal of the American Medical Association* 251: 2956.

NESCAUM. 2003. *Mercury Emissions from Coal-Fired Power Plants. The Case for Regulatory Action.* EPA Publication, 258pp.

Newman, M.C. and C.H. Jagoe. 1996. *Ecotoxicology: A Hierarchical Treatment.* CRC Press, Boca Raton, 412pp.

Newman, M.C. and M.A. Unger. 2003. *Fundamentals of Ecotoxicology.* Second edition. CRC Press, Boca Raton.

Nigel, B. 1994 and 2000. *Environmental Chemistry.* Second and Third edition. Wuerz Publishing Ltd., Winnepeg, 378pp. and 420pp.

Om Metaller. 1980. Second edition. Statens Naturvårdsverk, Stockholm, Sweden, 262pp.

Prost, R. 1995. *Contaminated Soil.* Institut National de la Recherche Agronomique, Paris, 525pp. +CD.

Schuurmann, G. and B. Markert. 1998. *Ecotoxicology.* John Wiley, New York, 902pp.

Schwartz, J. et al. 1986. Relationship between childhood blood lead level and stature. *Pediatrics* 77: 281–288.

Wetzel, R.G. 1983. *Limnology.* Second edition. Saunders College Publishing, New York, 828pp.

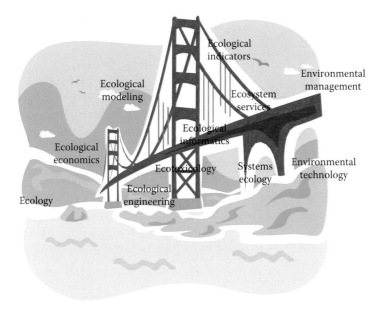

FIGURE 1.3
A conceptual bridge illustrating the close and integrated cooperation between subdisciplines of ecology and environmental management, which is a prerequisite for an up-to-date and holistic solution of the environmental problem.

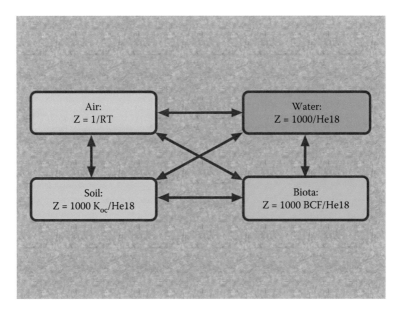

FIGURE 2.3
A fugacity model gives an overview of the distribution of chemical (toxic) compounds among the spheres: atmosphere, hydrosphere, lithosphere, and biosphere. The applied fugacity expressions are shown in the figure.

FIGURE 4.4
The opening screen image of EEP.

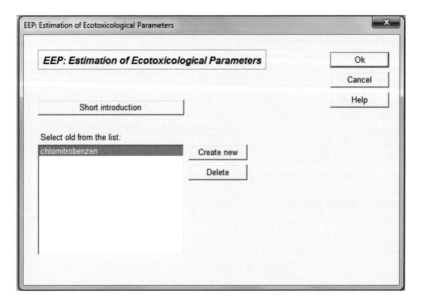

FIGURE 4.5
Either a previous applied compound or a new compound can be selected. In the continuous illustration of the application of EEP, we will use 1-chloro-4-nitro-benzene. It means we will select from the list of already-used compounds. For a new compound, the chemical name and the initial of the user must be given. When the compound has been selected, click OK and Figure 4.6 appears.

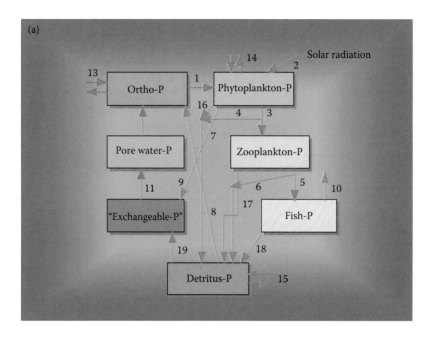

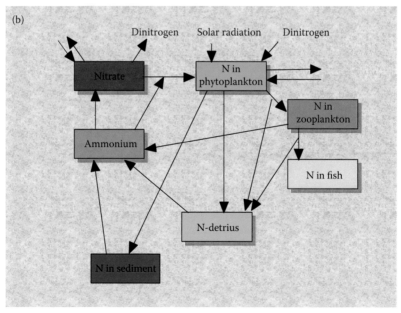

FIGURE 5.1
Phosphorus (a) and nitrogen (b) cycles in a lake by using a simple food chain nutrient–phytoplankton–zooplankton–fish to describe the processes of the cycle. The cycles are closed by mineralization of detritus and by release of nutrient by the sediment. (Reprinted from *Fundamentals of Ecological Modelling*, 3, Jørgensen, S.E. and G. Bendoricchio, 620 pp., Copyright (2001), with permission from Elsevier.)

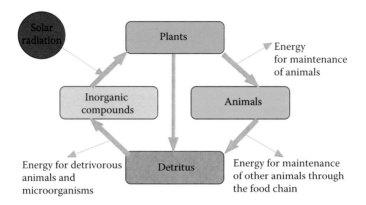

FIGURE 5.2
An ecosystem is a biochemical reactor that recycles matter. The input of energy is delivered by the solar radiation. The biologically important elements cycle and carry the energy that is utilized by heterotrophic organisms to support life processes.

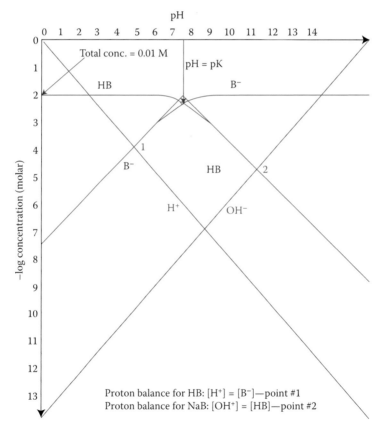

FIGURE 7.3
A double logarithmic diagram for an acid–base system with a total concentration of 0.01 m and pK = 6.0. The proton balance for HB and NaB is shown. The pH for the two cases are found as point 1 (pH = 4.0) and point 2 (pH = 9.0), respectively. Note that the total composition can be read at the diagram. B^- at point 1 is 0.0001 and HB at point 2 is 10^{-9}.

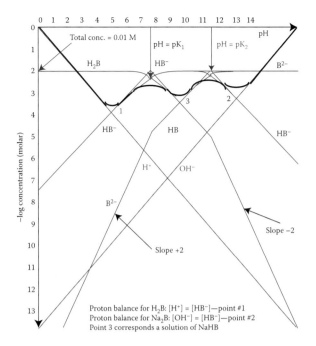

FIGURE 7.4
A double logarithmic diagram of the system H_2B, HB^-, and B^{2-}—total concentration = 0.01 M. Note that the slopes of the curves for H_2B and B^{2-} are +2 above pK_2 and below pK_1, respectively. The proton balance that was considered after approximations is shown for the two cases: 0.01 M H_2B and 0.01 M Na_2B. The composition of the two solutions can be read from the figure points 1 and 2, respectively. A 0.01 M solution of NaHB has the composition corresponding to point 3.

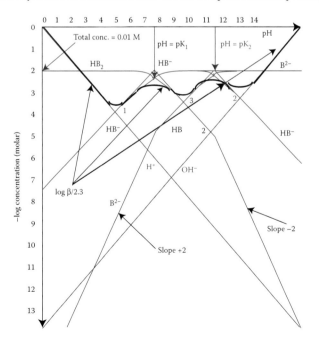

FIGURE 7.6
Log $(\beta/2.3)$ line for Figure 7.4. At points 1, 2, and 3, the line is 0.3 units above the intersections. At pK_1 and pK_2, log $(\beta/2.3)$ line is 0.3 below the intersection.

$$Me^{2+} \xrightarrow[\;(2)\;]{+L^-\quad K} MeL^+$$

$pe^o \;\Big\downarrow\Big\uparrow (1) \qquad\qquad (4)\Big\uparrow\Big\uparrow pe'o?$

$$Me^{3+} \xrightarrow[\;(3)\;]{+L^-\quad K'} MeL^{2+}$$

$n\,pe^o + \log K = \log K' + n\,pe^o$

When pe^o, $\log K$ and $\log K'$
are known, pe^o can be determined

FIGURE 7.25
It is possible to go from Me^{3+} to MeL^{2+} by the processes $(1) + (3)$ or by the processes $(2) + (4)$ and the two pathways must necessarily give the same result with respect to the equilibrium between Me^{3+} and MeL^{2+}. It implies that $n\,pe^0 + \log K = \log K' + n\,pe^{(0)}$.

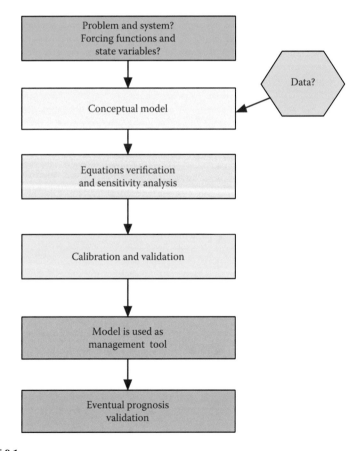

FIGURE 9.1
A tentative modeling procedure. (Adapted from Jørgensen, S.E. and B. Fath., *Fundamentals of Ecological Modelling*, Elsevier, Amsterdam, 2011.)

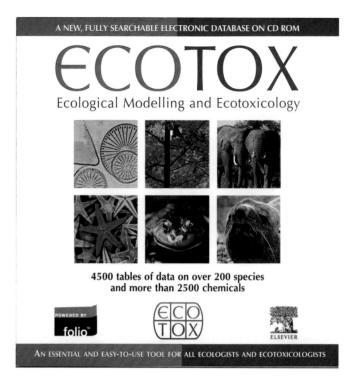

FIGURE 9.6
One of the major literature sources for parameter. (Jørgensen, L.A. et al. 2000. Ecotox. CD.)

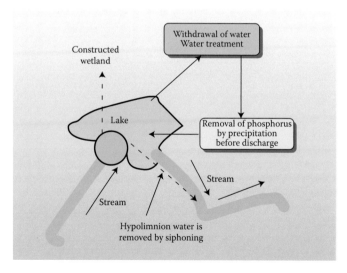

FIGURE 12.2
Control of lake eutrophication with a combination of chemical precipitation for phosphorus removal from wastewater (environmental technology), a wetland to remove nutrients from the inflow (ecotechnology, type 1 or 2), and siphoning off of hypolimnetic water, rich in nutrients and downstream (ecotechnology, type 3). The same combination of methods could, in principle, be applied to control the toxic substance contamination of a lake because precipitation can be used to remove toxic substances from wastewater, wetland can be used to reduce the toxic substance concentration of the inflowing stream, and siphoning can be used to remove toxic substances from the hypolimnetic water released from the sediment.

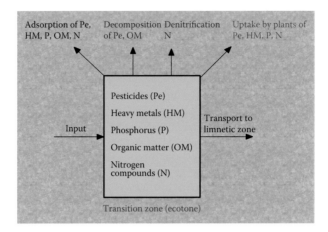

FIGURE 12.4
Figure showing that pesticides, heavy metals, phosphorus, organic matter, including veterinary medicine residues, and nitrogen compounds can be removed by wetlands. It is therefore very important to have a wetland area as a buffer zone between agriculture and ecosystems.

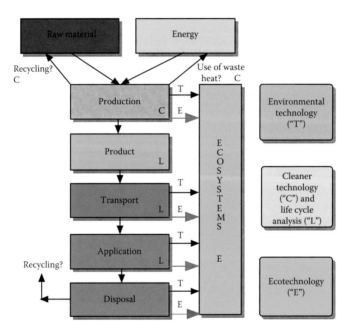

FIGURE 14.1
The arrows cover mass flows and the thin arrow indicates the control possibilities. Point pollution is indicated by a thin black arrow to the ecosystems and nonpoint pollution is indicated by a thick gray arrow. Production includes both industrial and agricultural production. The latter is most responsible for nonpoint pollution. Environmental technology (indicated by T) is mostly applied to solve the point pollution problems and ecotechnology (indicated by E). Cleaner technology (indicated by C) is investigating the possibilities to produce the products by another method that causes less environmental problems and may facilitate reuse and recycling. The diagram is based on a life cycle analysis—the product is followed from its production based on matter and energy to its final disposal as waste after its use. The diagram shows how the various solution tool bases are needed in an integrated way to solve all the problems associated with a product from its production to its disposal.

7

Calculations of Reactions and Equilibrium

7.1 Introduction

There are several million different chemical compounds known to us and 100,000 of these chemicals have environmental interest as they have industrial use and can be found in our everyday products. They are a threat to our environment and health (see Chapter 6). These compounds can, as all chemical compounds, participate in chemical reactions in the environment and are thereby transformed to other compounds. The properties of the compounds (see Chapter 3) determine the distribution in the spheres—lithosphere, hydrosphere, atmosphere, and technosphere. The ability to decompose has also been mentioned, and this process determines the time a compound remains in the environment. The decomposition can either be chemical or microbiological. This chapter mainly focuses on the chemical reactions in the environment of the emitted chemical compounds, including the chemical decomposition processes, but it also covers a few important physical processes that determine the transfer processes between the spheres. It is very important to perform calculations of these possible processes because they determine in which sphere and with which chemical compound we are dealing as a result of these transformation processes, and it determines thereby the environmental and health effect. To mention two illustrative examples: (1) It is important whether a toxic substance is in the atmosphere or in the lithosphere—just consider sulfur dioxide. (2) Heavy metals are able to form complexes, for instance, with chloride ions and the complexes have a completely different solubility, mobility, and effect (the toxic effect is usually much less). It is implicit that cadmium is much less toxic in a marine environment than in freshwater bodies. This chapter presents the chemical reaction calculations, where the equilibrium calculations are crucial. A method to use graphic representation to overview the different possible compounds by the equilibrium will be a core topic. It is possible to make the calculations without the graphs, but in this context it is preferable to use and demonstrate the graphic method because the results are given as very illustrative graphic overviews. Owing to the possibilities for fast calculations using computers, it is not necessary to use the graphic method, but the graphs present a good overview of the different components and their concentration. A few physical–chemical calculation methods that we have not yet covered in Chapters 3 and 4 are also included to ensure that we can answer the crucial question: Which components at which concentrations do we have and where in the environment? The calculations included in this chapter focus on the distribution among the atmosphere, hydrosphere, and lithosphere by the presentation of Henry's law and adsorption isotherms. In this context, we discuss the estimation methods in addition to what is covered in Chapter 4. Examples have been used throughout the chapter to give the best possible illustration of the calculation methods.

7.2 Equilibrium Constant

The number of possible reactions among the millions of chemicals is enormous, and it is of course not possible to set up a table of the equilibrium constants for all these reactions. We can, however, apply the standard free energies of formation of chemical compounds, ΔG_o. The standard free energy of formation of a compound is the free energy of reaction by which it is formed from its elements, when all the reactants and products are in the standard state, that is, the activities are all 1 or by ideal conditions, it means that the concentrations are all 1. Free energy equations can be added and subtracted just as thermochemical equations. It implies that the free energy of any reaction can be calculated from the sum of the free energies of the products minus the sum of the free energies of the reactants:

$$\Delta G_o = \sum G_o(\text{products}) - \sum G_o(\text{reactants}) \tag{7.1}$$

Free energy describes the chemical affinity under conditions of constant temperature and pressure: $\Delta G = G(\text{products}) - G(\text{reactants})$. When the free energy is zero, the system is in a state of thermodynamic equilibrium. When the chemical energy change is positive for a proposed process, the network must be put into the system to effect the reaction, otherwise it cannot take place. When the free energy change is negative, the reaction can proceed spontaneously by providing a useful network. As it has been shown in physical chemistry, the equilibrium constant K is related to $-\Delta G_o$ by the following equation; for further details, see Jørgensen (2012) and textbooks in physical chemistry:

$$-\Delta G_o = RT \ln K \tag{7.2}$$

If we consider the process $aA + bB = cC + dD$, we get

$$K = \frac{\{C\}^c\{D\}^d}{\{A\}^a\{B\}^b} \tag{7.3}$$

where {} indicates the fugacity or activity (if partial pressure is considered as a unit, the equilibrium constant is of indicated as K_p and if concentrations units are considered the equilibrium constant is indicated as K_c), in contrast to [], which indicates concentration. The equilibrium expression is also denoted as the mass law. The equilibrium constant may have different names corresponding to application of the mass law on different reactions. For instance, the equilibrium constant may be called the formation constant when $A + B$ forms AB, or if AB is a complex, the constant may be called a complexity constant or stability constant. The equilibrium constant is called dissociation constant when AB is dissociated into A^+ and B^-, and acidity constant when an acid HA is dissociated into a hydrogen ion and the corresponding base: $HA = H^+ + A^-$. The equilibrium constant for the opposite process is called a base constant. For a dissolution process, we talk about the solubility product.

EXAMPLE 7.1

A chemical plant discharges wastewater containing 26 g/L cyanide. The wastewater is treated by complete oxidation by which cyanide is oxidized to cyanate, using NaClO at pH > 10.5. After the oxidation, cyanate is hydrolyzed by the addition of acid whereby it is transformed to ammonium and carbon dioxide. The ΔG_o values by room temperature (25°C) are the following:

Cyanate (CNO^-)	-98.7 kJ/mol
Oxonium (H_3O^+)	-237.2 kJ/mol
Ammonium (NH_4^+)	-79.5 kJ/mol
Carbon acid (H_2CO_3)	-623.4 kJ/mol

A. Balance the two equations for the two reactions applied in the wastewater treatment.
B. Which pH value must be applied to ensure that the cyanate concentration after the hydrolysis is ≤ 0.43 µg/L?

The system is considered closed.

Solution

Oxidation:

$$CN^- + OCl^- \rightarrow CNO^- + Cl^- \tag{I}$$

Redox balance:

$$
\begin{array}{lll}
\text{C:} & 2 \rightarrow 4 & 2\uparrow \\
\text{Cl:} & 1 \rightarrow -1 & 2\downarrow
\end{array}
$$

Hydrolysis:

$$CNO^- + 2H_3O^+ \rightarrow NH_4^+ + H_2CO_3 \qquad \text{(II)}$$

K_{II} for (II):

$$\Delta G_{II^\circ} = \sum \Delta G_{\circ\,Product} - \sum G_{\circ\,Reactant}$$

$$= (-623.4 - 79.5 - (-98.7 + 2(-237.2)))\,kJ/mol$$

$$= 129.8\,kJ/mol$$

$$\log(K_{II}) = \frac{-\Delta G_{II}^\circ}{(RT\ln(10))} = \frac{129.8}{5.7} = 22.8$$

If it is assumed that pH is sufficiently low to ensure that the two result-ing compounds (see the reaction) are not dissociated, then

$$\left[NH_4^+\right]_{slut} \approx [H_2CO_3]_{slut} \approx [CN^-]_{start} = \frac{26\,g\ cyanide/L}{26\,g\ cyanide/mol} = 1\,M$$

$$[CNO^-]_{slut} = 0.43\,\mu g\ cyanate/L = 1\cdot 10^{-8}\,M$$

$$K_{II} = 10^{22.8} = \left[NH_4^+\right]\frac{[H_2CO_3]}{[CNO^-]}[H_3O^+]^2$$

$$[H_3O^+] = 10^{-7.4}\,M,\ at\ pH = 7.4$$

The assumption was with other words not correct, but it is assumed that all C(IV) is in the form of HCO_3^-, that is, $\left[HCO_3^-\right] \approx 1\,M$. A combina-tion of the equilibrium expression for process (II) and the equilibrium expression for carbon acid's protolysis yields

$$K_{s1}\cdot K_{II} = \frac{\left[NH_4^+\right]\left[HCO_3^-\right]}{[CNO^-][H_3O^+]}$$

By solving for the unknown, $[H_3O^+]$, we obtain pH = 8.46. HCO_3^- is dominating at this pH. Our assumption was now fully acceptable, as $pK_s = 9.25$ for NH_4^+.

7.3 Activities and Activity Coefficients

Any activity can be written as the product of concentration and activity coefficient: $\{A\} = q\,[A]$. The activity is basically defined in such a way that the activity coefficient $q = \{A\}/[A]$ approaches unity as the concentration of all solutes approaches zero. It means for a solvent as water the activity coefficient becomes unity as the solvent approaches the pure ionic medium, that is, when all concentrations other than the medium ions approach zero.

The activity coefficient, q, can be found for individual ions by empirical expressions as given in Table 7.1, where I is the ionic strength $I = 0.5 \sum C_i Z_i^2$ and Z_i = charge of the ion. A in the table is = $1.82 \times 10^6 (É^*T)^{2/3} \approx 0.5$ for water at room temperature. É is the dielectric constant. $B \approx 0.33$ for water at room temperature and a is an adjustable parameter corresponding to the size of the ion (see Table 7.2). Log q for ions is negative, which implies that q is less than 1 and decreases with increasing ionic strength and charge of the ion. The activity is less than the concentration because negative ions form a shield around a positive ion and positive ions form a shield around negative ions. The stronger the shield, the more ions the solution contains, and thus the higher the ionic strength. The electrical force is furthermore proportional to the charge in second, which at least explains that the effect is increasing more than proportional to the charge of the ion. The equations

TABLE 7.1

Equations for Individual Activity Coefficients

Name of the Approximation	Equation: Log q =	Valid at I <
Debye–Hückel	$-AZ^2\sqrt{I}$	0.005 M
Extended Debye–Hückel	$-AZ^2\sqrt{I}/(1 + Ba\sqrt{I})$	0.1 M
Güntelberg	$-AZ^2\sqrt{I}/(1 + \sqrt{I})$	0.1 M
Davies	$-AZ^2(\sqrt{I}/(1 + \sqrt{I}) - 0.2I)$	0.5 M

TABLE 7.2

Parameter a for Individual Ions

Ion Size Parameter a	For the Following Ions
9	H^+, Al^{3+}, Fe^{3+}, La^{3+}, Ce^{3+}
8	Mg^{2+}, Be^{2+}
6	Ca^{2+}, Zn^{2+}, Cu^{2+}, Sn^{2+}, Mn^{2+}, Fe^{2+}
5	Ba^{2+}, Sr^{2+}, Pb^{2+}, CO_3^{2-}
4	Na^+, HCO_3^-, $H_2PO_4^-$, acetate, SO_4^{2-}, HPO_4^{2-}, PO_4^{3-}
3	K^+, Ag^+, NH_4^+, OH^-, Cl^-, ClO_4^-, NO_3^-, I^-, HS^-

in Table 7.1 can therefore be understood as a consequence of the electrical forces in solutions. They cannot however be proved, but are useful empirical correlations.

It is clear from Section 7.2 that we can find the equilibrium constant that presumes activities or fugacities by the application of Equations 7.1 through 7.3. Equilibrium constants taken from handbook tables are also based on activities and fugacities, while we are most often interested in concentrations. Introduction of the activity coefficient, q, makes it possible, however, to set up the following relationships:

$$K = \frac{\{C\}^c\{D\}^d}{\{A\}^a\{B\}^b} = \left(\frac{[C]^c[D]^d}{[A]^a[B]^b}\right)\left(\frac{q_C^c q_D^d}{q_A^a q_B^b}\right) \tag{7.4}$$

Concentrations can now be determined, provided the activity coefficients are known.

7.4 Mixed Equilibrium Constant

Usually, many equilibrium calculations are carried out for the same solution with a well-defined ionic strength. It would therefore be beneficial to find an equilibrium constant K′ valid for concentrations for the considered solution. From Equation 7.4, we obtain

$$K' = \frac{K q_A^a q_B^b}{q_C^c q_D^d} \tag{7.5}$$

In accordance with the IUPAC's convention for determination of pH, we should consider $pH = -\log \{H^+\}$. It is therefore suggested to use a so-called mixed acidity constant, K'_{am}, which can be found from K_a for the process $HA \Leftrightarrow A^- + H^+$:

$$
\begin{aligned}
K_a &= \frac{\{A^-\}\{H^+\}}{\{HA\}} = \frac{[A^-]q_{A-}\{H^+\}}{[HA]\, q_{HA}} \\
&= \frac{K'_{am}\, q_{A-}}{q_{HA}} = \frac{K'_a\, q_{A-}\, q_{H+}}{q_{HA}}
\end{aligned}
\tag{7.6}
$$

q_{HA} is of course 1.0, if HA has no charge (see the equations in Table 7.1). Calculations of pH presumes the application of K'_{am}. In environmental calculations, where the accuracy is generally lower than that for laboratory calculations, it is recommended to apply activity coefficients for salinities above 0.1%, whereas it is normally not necessary to apply activity coefficients for aquatic ecosystems with salinity below 0.1%.

EXAMPLE 7.2

Find the mixed acidity constant and the concentration quotient (equilibrium constant for all dissolved carbon dioxide and for the ammonium ion in marine environment with a salinity of 2.6% (assume that it is sodium chloride), when logarithm to the acidity constant in distilled water for all dissolved carbon dioxide at the actual temperature is known to be 6.2 and for the ammonium ion at the actual temperature is 9.2.

Solution

$$I = 0.5 \left(\frac{26}{(23 + 35.5)} \right) + \left(\frac{26}{(23 + 35.5)} \right) = 0.445 \text{ M}$$

Davies equation is applied.

For the hydrogen carbonate ion, the ammonium ion, and the hydrogen ion:

$\log q = -0.5 \ (\sqrt{0.445}/(1 + \sqrt{0.445}) - 0.2 \cdot 0.445) = -0.156$
$q = 0.698 \approx 0.7$

The acidity constants for all dissolved carbon dioxide:

$\log K_a' = \log K_a - 2 \log q = 6.2 + 0.312 \approx 6.5$
$\log K_{am}' = 6.2 - \log q = 6.2 - (-0.156) \approx 6$

The acidity constant for the ammonium ion:

$\log K_a' = \log K_a + \log q - \log q = 9.2$
$\log K_{ám} = \log K_a + \log q = 9.2 + (-0.156) = 0.9044 \approx 9.0$

A summary of the calculations:

1. Find the ionic strength as strength $I = 0.5 \Sigma C_i Z_i^2$.
2. Find the activity coefficient by one of the equations in Table 7.1; for instance, Davies equation
   $-AZ^2 \left(\sqrt{I}/(1 + \sqrt{I}) - 0.2I \right)$ provided I = or <0.5 M.
3. It is now possible to find the equilibrium constants based on concentrations but for the hydrogen ion (pH) the activity by Equations 7.5 and 7.6.

7.5 Classification of Chemical Processes and Their Equilibrium Constants

Chemical processes may be divided into four classes all of which occur very frequently in all aquatic ecosystems and in soil water and sediment water solutions:

I. *Acid–base reactions*: These are processes characterized by a transfer of a proton. Acids are hydrogen ion donors and bases are hydrogen ion acceptors:

$$HA \Leftrightarrow A^- + H^+ \qquad (7.7)$$

HA is therefore an acid and A^- a base. $HA - A^-$ is denoted as an acid–base pair. This is called a half reaction because the hydrogen ion cannot under normal chemical conditions exist alone and will inevitably be taken up by another component that is a base:

$$HA + B^- \Leftrightarrow A^- + HB \qquad (7.8)$$

Water is both an acid and a base—it is an ampholyte—and can therefore react with both acids and bases:

$$HA + H_2O \Leftrightarrow A^- + H_3O^+ \qquad (7.9)$$

H_3O^+ is called the oxonium ion. The equilibrium constant for process (7.7) is called the acidity constant. The acidity constants listed in most handbooks of chemistry are, however, the acidity constant for process (7.9), but we use it as it was the equilibrium constant for process (7.7), corresponding to the concentration of water in water often implicitly included in the equilibrium constant for aquatic solutions. Furthermore, we often do not distinguish between hydrogen ions and oxonium ions because hydrogen ions in solution will anyhow "carry" a water molecule.

When the process is that the base A^- takes up a hydrogen ion, the equilibrium constant is called a base constant, K_b.

$$\text{As seen } K_b = \frac{1}{K_a} \qquad (7.10)$$

The equilibrium constant for process (7.8) can easily be found from the two acidity constants for HA, K_{aA} and HB, K_{aB}

$$\text{for (7.8)} = \frac{\{A^-\}\{HB\}}{\{HA\}\{B^-\}} = \frac{K_{aA}}{K_{aB}} \qquad (7.11)$$

The acidity constant for water is

$$\frac{\{H^+\}\{OH^-\}}{\{H_2O\}} = 10^{-15.74} \quad \text{(room temperature)} \tag{7.12}$$

$\{H_2O\} \approx [H_2O] = 1000/18 = 55.56$ because water has no charge. It implies that what is called water's ionic product:

$$K_w = \{H^+\}\{OH^-\} = 10^{-14.00} \quad \text{(at room temperature)}$$

II. *Precipitation and dissolution*: These are reactions characterized by a change in solubility. Dissolution and precipitation processes are generally slower than reactions among dissolved species. Electrolytes may dissolve according to the following reaction:

$$A_m B_n(s) \Leftrightarrow mA^{n+} + nB^{m-} \tag{7.13}$$

The opposite process is the corresponding precipitation process. If the equilibrium expression is used on process (7.13), we obtain the solubility product

$$K_s = \{A^{n+}\}^m \cdot \{B^{m-}\}^n \tag{7.14}$$

III. *Complex formation*: It is a reaction by which two or more components form (more complex) compounds. This reaction type is particularly known for a reaction between metal ions (named central atom) and various organic compounds (named ligands). Frequently, more than one ligand can be complex bound to the central atoms, for instance,

$$M + L = ML \quad K_1 = \frac{\{ML\}}{\{M\}\{L\}} \tag{7.15}$$

$$ML + L = ML_2 \quad K_2 = \frac{\{ML_2\}}{\{ML\}\{L\}} \tag{7.16}$$

$$ML_2 + L = ML_3 \quad K_3 = \frac{\{ML_3\}}{\{ML_2\}\{L\}} \tag{7.17}$$

$$ML_i + L = ML_{i+1} \quad K_i = \frac{\{ML_{i+1}\}}{\{ML_i\}\{L\}} \tag{7.18}$$

The process whereby two or more ligands react simultaneously with the central atom is of course also possible:

$$M + L_2 = ML_2 \quad \beta_2 = K_2 K_1 = \frac{\{ML_2\}}{\{M\}\{L\}^2} \tag{7.19}$$

As seen, the equilibrium constant where i ligands are simultaneously reacting with the central atom is denoted by $\beta_1 = K_1 K_2 K_3...K_i$.

Reactions can also take place by the addition of protonated ligands:

$$M + HL = ML + H^+ \quad ^*K_1 = \frac{\{ML\}\{H^+\}}{\{M\}\{HL\}} \tag{7.20}$$

A parallel expression is used for the reaction with the second HL, third HL, and so on. By multiplying {L} in the nominator and denominator of the expression in Equation 5.20, it is seen that $^*K_1 = K_1 K_{aL}$. Note that we have used the more generally expressed rule that the equilibrium constant for a process, K^*, that consists of i steps, is equal to the product of the i equilibrium constants of the steps, $K_1, K_2, K_3, \dots, K_i$:

$$^*K = K_1, K_2, K_3, \dots, K_i \tag{7.21}$$

or

$$\log {^*K} = \log K_1 + \log K_2 + \log K_3 + \cdots + \log K_i \tag{7.22}$$

IV. *Redox reactions*: These are processes characterized by a transfer of electrons.

Reductants are electron donors and oxidants are electrons acceptors. The mass law may of course be also applied on redox processes, for instance (e denotes the electron with one negative charge),

$$Fe^{3+} + e = Fe^{2+} \tag{7.23}$$

$$K = \frac{\{Fe_{2+}\}}{\{Fe^{3+}\}\{e\}}$$

This process is similar to what was mentioned under acid–base reactions, called a half reaction. Free electrons do not exist. They are inevitably taken up by another electron acceptor. For the process (7.23), it is possible to determine an equilibrium constant. Log K can in handbooks be found to be 12.53, but the realization of the process requires a coupling to another half reaction, for instance, oxygen:

$$O_2 + 4H^+ + 4e = 2H_2O \tag{7.24}$$

with the equilibrium expression

$$K_o = \frac{1}{p_{O_2}} \{H^+\}^4 \{e\}^4 \tag{7.25}$$

$$\log K_o = 83.1$$

It is possible to couple the two processes to a feasible redox process:

$$O_2 + 4H^+ + 4Fe^{2+} = 2H_2O + 4Fe^{3+} \tag{7.26}$$

The equilibrium constant for this process, K_r, is found from the equilibrium constants of the two half reactions:

$$K_r = \frac{K_o}{K^4} \tag{7.27}$$

$$\log K_r = \log K_o - 4\log K = 83.1 - 4 \cdot 13.0 = 31.1$$

In environmental chemistry, it is difficult to separate the many reactions that can take place simultaneously. We are therefore forced in aquatic environmental chemistry to make equilibrium calculations of several processes simultaneously. It will be shown later in this chapter how it is possible to overview many processes simultaneously, particularly to assess which of the many simultaneous processes are of importance and which are negligible. In addition, it may often be advantageous to apply what is called a conditional equilibrium constant. It is an equilibrium constant that is only valid under given conditions, for instance, that pH has a certain value or chloride has a certain concentration (of interest for a marine environment). It is, in other words, conditional that certain components have a constant or almost constant concentration. Many aquatic ecosystems have a stable pH at least for a short period. If water with pH = 7.0 is considered and the partial pressure of oxygen is 0.21 atmosphere, we get for process (7.26) the following equilibrium expression, replacing activities by concentrations:

$$K_r = \frac{[Fe^{3+}]^4}{0.21 \cdot 10^{-28} \cdot [Fe^{2+}]^4} = 10^{31.1}$$

The ratio $[Fe^{3+}]/[Fe^{2+}]$ is therefore at pH = 7.0 and at equilibrium with oxygen in the atmosphere ≈ 4.0.

By incorporating $0.21 \cdot 10^{-28}$ into the equilibrium constant, we get a conditional equilibrium constant:

$$\frac{[Fe^{3+}]}{[Fe^{2+}]} = 4.0$$

which is valid only under the condition that pH = 7.0 and that the aquatic ecosystem is in equilibrium with the oxygen in the atmosphere.

7.6 Many Simultaneous Reactions

Many processes occur simultaneously in aquatic ecosystems. Later in this chapter, we will treat the four types of processes presented above and include in this presentation how to provide an overview of many simultaneous processes of the *same* reaction type. The overview makes it possible to distinguish between processes of importance and processes that are negligible in the context. The double logarithmic presentation will be presented as an excellent tool to overview the processes, which is very important in the management of aquatic ecosystems, as the chemical processes determine the conditions for the biological processes in the aquatic ecosystems. An integrated environmental management requires that the physical, chemical, and biological processes and their interactions are considered simultaneously, as they determine the conditions for all possible processes.

7.7 Henry's Law

An increase or decrease in the concentration of components or elements in ecosystems are of vital interest, but the observation of trends in global changes of concentrations might be even more important as they may cause changes in the life conditions on Earth. The concentrations in the four spheres—atmosphere, lithosphere, hydrosphere, and biosphere—are of importance in this context. They are determined by the transfer processes and the equilibrium concentrations among the four spheres. The solubility of a gas at a given concentration in the atmosphere can be expressed by means of Henry's law, which determines the distribution between the atmosphere and the hydrosphere:

$$p = H^* x \tag{7.28}$$

where
   p = the partial pressure
   H = Henry's constant
   x = molar fraction in solution

   H is dependent on temperature (see Table 7.3), and H is expressed (usually) in atmospheres. It may be converted to pascals, as 1 atmosphere = 101,400 Pa. A dimensionless Henry's constant may also be applied. As $p = RT\,n/v = RTc_a$ and $x = c_h/(c_h + c_w)$, where c_a is the molar concentration in the atmosphere of component h, expressed in mol/L and c_h is the concentration in the hydrosphere expressed also in mol/L and c_w is the mol/L of water (and other

TABLE 7.3

Henry's Constant (atm) for Gases as a Function of Temperature

Gas	Temperature (°C)						
	0	5	10	15	20	25	30
Acetylene	0.72	0.84	0.96	1.08	1.21	1.33	1.46
Air (atm)	0.43	0.49	0.55	0.61	0.66	0.72	0.77
Carbon dioxide	73	88	104	122	142	164	186
Carbon monoxide	0.35	0.40	0.44	0.49	0.54	0.58	0.62
Hydrogen	0.58	0.61	0.64	0.66	0.68	0.70	0.73
Ethane	0.13	0.16	0.19	0.23	0.26	0.30	0.34
Hydrogen sulfide	26.80	31.50	36.70	42.3	48.30	54.50	60.90
Methane	0.22	0.26	0.30	0.34	0.38	0.41	0.45
Nitrous oxide	0.17	0.19	0.22	0.24	0.27	0.29	0.30
Nitrogen	0.53	0.60	0.67	0.74	0.80	0.87	0.92
Nitric oxide	–	1.17	1.41	1.66	1.98	2.25	2.59
Oxygen	0.25	0.29	0.33	0.36	0.40	0.44	0.48

Source: Jørgensen, S.E., S. Nors Nielsen, and L.A. Jørgensen. 1991. *Handbook of Ecological Parameters and Ecotoxicology*. Elsevier, Amsterdam. Published as CD under the name ECOTOX, with L.A. Jørgensen as first editor in year 2000.

Note: The values in the table are Henry's constant ($\times 10^{-5}$).

possible components). If we consider only two components in the hydrosphere, h and water, and that $c_h \ll c_w$, we can replace $(c_h + c_w)$ with the concentration of water in water $= 1000/18 = 55.56$ mol/L. According to these approximations, we obtain the following equation:

$$\frac{c_a}{c_h} = \frac{H}{(R \times T \times 55.56)} \tag{7.29}$$

where $H/(R \times T \times 55.56)$ is the dimensionless Henry's constant.

In aquatic environmental chemistry, we often know the partial pressure in atm, p_a, and want to calculate the concentration in water. In this case, we often use the following expression:

$$C_h = K_H p_a \tag{7.30}$$

K_H is a constant $= 55.56/H$. For instance, if we use the values in Table 7.3 for carbon dioxide, we get that K_H at 20°C is $55.56/1.42 \cdot 10^5 = 3.91 \cdot 10^{-2} = 10^{-1.41}$. In the year 2016, the partial pressure of carbon dioxide in the atmosphere is close to 0.0004 atm, which corresponds at 20°C to a concentration of $3.91 \cdot 10^{-2} \cdot 0.0004 = 0.0000156$ M in an aquatic ecosystem in equilibrium with the atmosphere.

EXAMPLE 7.3

The solubility of oxygen in freshwater is 11.3 mg/L at 10°C. Show that it corresponds to Henry's constant found in Table 7.3 for oxygen.

Solution

Henry's law is applied to find the molar fraction in water, x (the partial pressure of oxygen is 0.21 atm, corresponding to 21% oxygen in the atmosphere):
$x = 0.21/33000 = 6.36 \cdot 10^{-6}$. This is translated to mg/L:
Oxygen dissolved mg/L $= 6.36 \times 10^{-6} \cdot 55.56 \times 32 \times 1000 = 11.31$ mg/L.

7.8 Adsorption

Water is in contact with suspended solid or sediment and the equilibrium between solid and water is therefore very important for the water quality. The soil–water distribution may be expressed by one of the following two adsorption isotherms (see also Chapter 3, Equations 3.1 and 3.2):

$$a = kc^{b} \tag{7.31}$$

$$a = \frac{k'c}{c + b'} \tag{7.32}$$

where a is the concentration in soil, c is the concentration in water, and k, k', b, and b' are constants. Equation 7.31 corresponds to Freundlich adsorption isotherm and is a straight line with slope b in a log–log diagram, since log a = log k + b log c. This is shown in Figure 7.1.

Equation 7.32, the Langmuir adsorption isotherm, is an expression similar to Michaelis–Menten equation. If 1/a is plotted versus 1/c (see Figure 7.2), we obtain a straight line, the so-called Lineweaver–Burk plot, as 1/a = 1/k' + b'/k'c. When 1/a = 0, 1/c = –1/b' and when 1/c = 0, 1/a = 1/k'.

b is often close to 1 and c is small for most environmental problems. This implies that the two adsorption isotherms get close to a/c = k, and k becomes a distribution coefficient. k for 100% organic carbon, usually denoted K_{oc}, may be *estimated* from K_{ow}, the octanol–water distribution coefficient, which is the solubility in octanol divided by the solubility in water. For more details about these properties of chemical compounds, see Chapter 3. Several estimation equations have been published in the literature; see Chapter 4 and Jørgensen et al. (1997). The following log–log relationships between K_{oc} (100% organic carbon presumed) and K_{ow} are typical examples (compare with Chapter 4):

$$\log K_{oc} = -0.006 + 0.937 \log K_{ow} \quad \text{(Jørgensen, 2000)} \tag{7.33}$$

$$\log K_{oc} = -0.35 + 0.99 \log K_{ow} \quad \text{(Leeuwen and Hermens, 1995)} \tag{7.34}$$

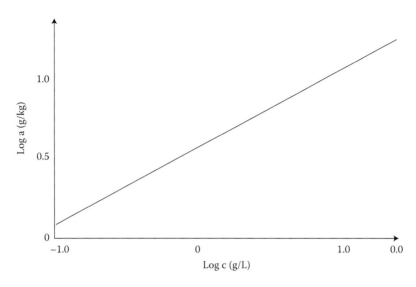

FIGURE 7.1
A log–log plot of the Freundlich adsorption isotherm. The slope, which is $1.15/3 = 0.383$, represents b in Equation 6.4 and $\log k = 0.48$, which means that $k = 3.1$. The equation for the plot shown is therefore $a = 3.1c^{0.383}$.

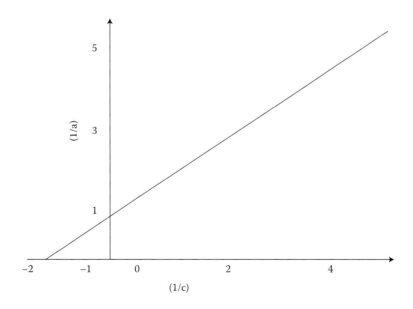

FIGURE 7.2
A Lineweaver–Burk plot where $1/a$ is plotted versus $1/c$. From the plot, it is possible to read that $1/b' = -(-1.5)$, or $b' = 2/3$ and that $k' = 1$ ($1/k' = 1$). This means that t is a Langmuir adsorption isotherm: $a = c/(c + 2/3)$.

In the case that the carbon fraction of organic carbon in soil is f, the distribution coefficient, K_D, for the ratio of the concentration in soil and in water can be found as $K_D = K_{oc} \cdot f$.

If the solid is an activated sludge (from a biological treatment plant) instead of soil, K_D can be found as described above or the equation log $K_D = 0.39 + 0.67 \log K_{ow} \cdot K_{ow}$ can be found for many compounds in the literature (see also Chapter 4), but if the solubility in water is known, it is possible to estimate the partition coefficient *n*-octanol–water at room temperature by use of a correlation between water solubility in μmol/L and K_{ow}. A graph of this relationship is shown in Figure 4.1.

7.9 Double Logarithmic Diagrams Applied on Acid–Base Reactions

The chemistry of the aquatic environment always involves many simultaneous reactions, which can be difficult to overview unless we use computer programs or double logarithmic diagrams, which allow us quickly to assess the approximate concentrations. It includes a determination of which components we do not need to consider because they have relatively very small concentrations. The double logarithmic diagram is applied in this handbook because it is relatively easy to construct and is very illustrative. The double logarithmic diagram for acid–base reactions plots the logarithmic of the concentration of the various species versus pH.

If we consider an acid–base reaction: $HA \Leftrightarrow A^- + H^+$, the logarithmic form of the equilibrium constant expression is

$$pK = -\log\{A^-\} - \log\{H^+\} + \log\{HA\} = -\log\{A^-\} + pH + \log\{HA\} \quad (7.35)$$

or in the form of the so-called Henderson–Hasselbach equation:

$$pH = pK + \frac{\log\{A^-\}}{\{HA\}} \quad (7.36)$$

Note that "p" is a general abbreviation for "−log." K is the equilibrium constant.

From Equation 7.36, it is clear that $pH \ll pK$ when $\{HA\} \gg \{A^-\}$. As the acid–base system only has the two forms HA and A^-, $\{HA\} \approx [HA] \approx C$, where C is the total concentration of the acid–base system. Log $\{A^-\}$ at low pH values can be derived from Equation 7.35: $\log\{A^-\} = pH - pK + C \approx \log[A^-]$. This will correspond to a straight line with the slope + pH in a logarithmic diagram. For pH = pK, the line will take the value C. A straight line through the point (pK, C) with a slope of +1, represents therefore log [A⁻] in a diagram where log c_i (c_i symbolizing various species) versus pH.

At $pH \gg pK$, $\{A^-\} \approx [A^-] \approx C$. Log $\{HA\} = pK - pH + C \approx \log [HA]$. This implies that log [HA] in a double logarithmic diagram for high pH values is represented by a straight line with the slope –1 and going through the point (pK, C).

For $pH = pK$, we know from Equation 7.36 that $\{HA\} = \{A^-\} = C/2 \approx [HA] = [A^-]$. The following table summarizes these results.

	Log [HA]	Log [A⁻]
$pH \ll pK$	$\log C$	$pH - pK + C$
$pH = pK$	$\log (C/2)$	$\log (C/2)$
$pH \gg pK$	$pK - pH + C$	$\log C$

Figure 7.3 shows a double logarithmic diagram for a 0.01 M acid HB (log C = –2) with pK = 6.0. The diagram is drawn by the use of straight lines for log [HB] and log [B⁻] at low and high pH, and the point (pK, C/2) that is valid for both log [HB] and log [B⁻]. The gap between low pH and pH = pK can easily be drawn and correspondingly for the gap from pH = pK and high pH.

The double logarithmic diagram represents two equations: the mass equation expression and the information that the total concentration is 0.01 M. We have, however, four unknowns: [HB], [B⁻], [H⁺], and [OH⁻]. We therefore need two more equations. Water's ion product $[H^+][OH^-] = 10^{-14}$ or on logarithmic form: $\log [H^+] + \log [OH^-] = -14 = pK_w$ can be used as the third equation. The fourth and last equation is the information that a solution will always be uncharged—that the sum of the concentrations of positive charged ions times their charge = the sum of the concentrations of the negative charged ions times their charge. In most cases it is, however, more beneficial to use that the concentrations resulting from dissociation of hydrogen ions is equal to the concentrations of ions that have taken up hydrogen ions. It is called the proton balance. To use this fourth equation to assess the composition of a solution included pH, it is recommendable to write the components that are in the solution before any reaction.

For instance, if an acid HB in water is considered, the components before any reaction is HB and H₂O. These two components can therefore not result from dissociation or uptake of hydrogen ions. Dissociation of hydrogen ions from these two components can only result in the formation of B⁻ and OH⁻ and uptake of hydrogen ions is only possible for water forming the oxonium ions, which is often just written as H⁺ only. The proton balance therefore yields the following equation:

$$[B^-] + [OH^-] = [H^+] \tag{7.37}$$

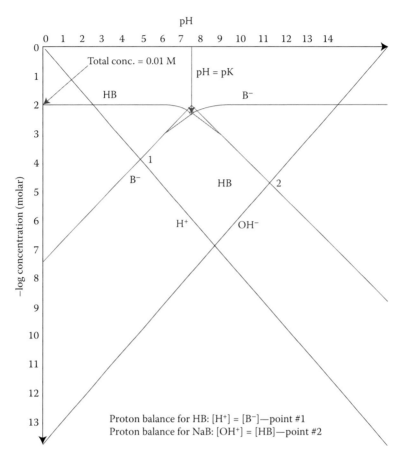

FIGURE 7.3

(**See color insert.**) A double logarithmic diagram for an acid–base system with a total concentration of 0.01 m and pK = 6.0. The proton balance for HB and NaB is shown. The pH for the two cases are found as point 1 (pH = 4.0) and point 2 (pH = 9.0), respectively. Note that the total composition can be read at the diagram. B^- at point 1 is 0.0001 and HB at point 2 is 10^{-9}.

As we have an acidic solution, pH is relatively low, and it is therefore assumed—at least in the first hand—that $+[OH^-] \approx 0$. Therefore, the proton balance gives

$$[B^-] = [H^+] \tag{7.38}$$

This equation corresponds to point 1 in Figure 7.3. This point represents the composition of 0.01 M HB solution. It is of course necessary to check the assumption that the hydroxide ions are negligible. As seen $[OH^-] = 10^{-9}$ at point 1. It was therefore fully acceptable to consider $[OH^-]$ as negligible compared with $[B^-]$ and $[H^+]$.

With the same argument it can be found that the proton balance for a solution of NaB yields $[OH^-] = [HB]$ corresponding to point 2 in Figure 7.4.

Figure 7.4 shows a double logarithmic diagram of the system H_2B, HB^- and B^{2-}. The total concentration is 0.01 M. In this case, the acid is able to dissociate two hydrogen ions and therefore has two pK values: pK_1 and pK_2. The H_2B curve will at $pH \geq pK_2$ have the slope -2, as seen, because the actual process is $H_2B \Leftrightarrow B^{2-} + 2H^+$. B^{2-} will correspondingly have the slope $+2$ at $pH \geq pK_1$.

The proton balance for the ampholyte NaHB will with good approximations yield the equation: $[H_2B] = [B^{2-}]$ as $[OH^-]$ and $[H^+]$ both $\ll [H_2B] = [B^{2-}]$. Point 3 in Figure 7.4 corresponds to the composition of the various species

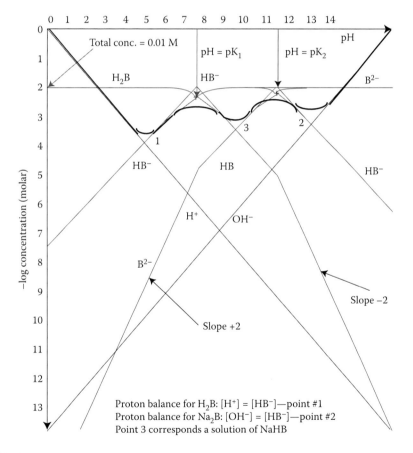

FIGURE 7.4
(See color insert.) A double logarithmic diagram of the system H_2B, HB^-, and B^{2-}—total concentration = 0.01 M. Note that the slopes of the curves for H_2B and B^{2-} are +2 above pK_2 and below pK_1, respectively. The proton balance that was considered after approximations is shown for the two cases: 0.01 M H_2B and 0.01 M Na_2B. The composition of the two solutions can be read from the figure points 1 and 2, respectively. A 0.01 M solution of NaHB has the composition corresponding to point 3.

in a 0.01 M NaHB solution with the pK values as shown in Figure 7.4, that is, 6.1 and 9.5. Point 1 corresponds to a 0.01 M H_2B solution, where the concentration of B^{2-} and hydroxide ions are considered negligible in the proton balance. The approximated proton balance is shown in the figure. Point 2 in Figure 7.4 corresponds to a 0.01 M B^2 solution. The proton balance is shown in the figure. H_2B and hydrogen ions are in this case considered negligible.

Phosphoric acid is a medium strong acid and therefore has a low pK_1 value, namely, 2.12. This implies that point 1 corresponding to the composition of a 0.4 M phosphoric acid solution yields an undissociated phosphoric acid concentration different from the total concentration where a weak acid solution clearly gives an undissociated acid close to the total concentration. If in the case of phosphoric acid, we have to read the concentration of phosphoric acid in a 0.4 M solution, it is advantageous to read the concentration of dihydrogen phosphate and deduct it from the total concentration. The concentration of dihydrogen phosphate is found from the diagram to be antilog (–1.4) corresponding to a concentration of about 0.04 M. The pH is 1.4. The concentration of phosphoric acid therefore becomes $0.4 - 0.04 = 0.36$ M.

7.10 Molar Fraction, Alkalinity, and Buffer Capacity

Introduction of molar fractions makes it possible to set up equations to compute the composition of acid–base solutions. As mentioned above, the application of a double logarithmic diagram corresponds to find x unknown concentrations from x equations. It is therefore of course possible to find a composition of a complex acid–base mixture by solving the equations. It is however in most cases faster and sufficiently accurate to use a double logarithmic diagram. Molar fractions may also be used in the double logarithmic diagram instead of concentrations. The concentrations for a number of different cases (different total concentrations) of the same components can be easily found by multiplying the molar fractions found in the diagram with the total concentration.

The molar fractions for the system H_2B, HB^-, and B^{2-} are shown in Figure 7.5. The three shown equations are found from the two mass equations and from the equations that expresses that the sum of the three molar fractions is one.

The buffer capacity, β, is defined as

$$\beta = \frac{dC}{dpH} \tag{7.39}$$

Acids and bases—important equations:

Molar fraction $H_2B = 1/(1 + K_1[H^+] + K_1K_2[H^+]^2)$

Molar fraction $HB^- = 1/(1 + [H^+]/K_1 + K_2/[H^+])$

Molar fraction $B^{2-} = 1/(1 + [H^+]/K_2 + [H^+]^2/K_1K_2)$

β = buffer capacity = $2.3 ([H^+] + [OH^-] + [HA][A^-]/([HA] + [A^-])$

$[Alk] = [HCO_3^-] + 2[CO_3^{2-}] - [H^+] + [OH^-] + \Sigma$ other base ions

NB!! Alkalinity is conserved.

FIGURE 7.5
Important definitions: molar fraction, buffer capacity, and alkalinity.

where dC is the strong acid or base added to the considered solution and dpH is the corresponding change in pH. When β is high, relatively much acid or base is needed to change the pH, whereas a low β value indicates that the pH is changed by the addition of a minor amount of acid or base. It can be shown by differentiation according to the definition given in Equation 7.39 that

$$\beta = 2.3 ([H^+]) + [OH^-] + \frac{[HA][A^-]}{([HA]+[A^-])} \tag{7.40}$$

β can be found directly from this expression. The various concentrations can be found by calculations or from the double logarithmic diagram. It is also possible to find and draw on the double logarithmic diagram the equation $\log \beta/2.3 = \log (([H^+] + [OH^-] + [HA][A^-]/([HA] + [A^-]))$. At very low pH, $[H^+]$ is dominating the expression (7.40) and a $\log (\beta/2.3)$ line in the double logarithmic diagram will therefore follow the line for $\log [H^+]$. At slightly higher pH, where $[H^+] = [A^-]$ and $[HA] \approx C = [HA] + [A^-]$, $\log \beta/2.3 = \log (2[H^+]) = \log (2[A^-]) = 0.3 + \log ([H^+]) = 0.3 + \log [A^-])$. Where the two lines for $[H^+]$ and $\log ([A^-])$ intersect, $\log (\beta/2.3)$ value will therefore be 0.3 above the intersection. At pH = pK, the β-expression is dominated by $[HA][A^-]/([HA] + [A^-] = C/2·C/2/C = C/4$. It means that the $\log (\beta/2.3)$ value will be 0.3 below the intersection of $[HA]$ and $[A^-]$ at pH = pK. At high pH, the expression is dominated by $[OH^-]$ and $\log [\beta/2.3]$ will therefore follow the line of $\log [OH^-]$, where $\log [OH^-] = \log [HA]$, $\log [\beta/2.3] = 0.3 + \log [OH^-] = 0.3 + \log [HA]$.

Figure 7.6 shows how a $\log (\beta/2.3)$ plot is found for the acid–base system in Figure 7.4. The double logarithmic diagram is applied to find which concentrations we have to consider in the buffer equation and which concentrations we can eliminate because they are negligible.

The two most important acidic components in the oceans are the carbon dioxide system and boric acid. As the pH in the oceans is 8.1, the buffer

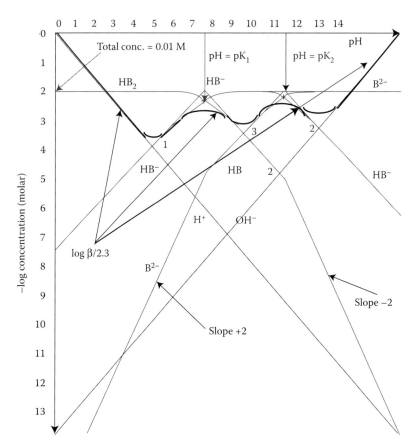

FIGURE 7.6
(**See color insert.**) Log ($\beta/2.3$) line for Figure 7.4. At points 1, 2, and 3, the line is 0.3 units above the intersections. At pK_1 and pK_2, log ($\beta/2.3$) line is 0.3 below the intersection.

capacity of water in the oceans can easily be found. The buffer capacity of the oceans is, however, much higher than the value obtained from the buffer capacities of the ions due their content of suspended clay minerals, which are able to buffer due to the following reaction:

$$3Al_2Si_2O_5(OH)_2 + 4SiO_2 + 2K^+ + 2Ca^{2+} + 12H_2O$$

$$\leftrightarrow 2KCaAl_3Si_5O_{16}(H_2O)_6 + 6H^+ \tag{7.41}$$

The pH dependence is indicated by the corresponding equilibrium expression in logarithmic form:

$$\log K = 6\log(H^+) - 2\log K^+ - 2\log Ca^{2+} \tag{7.42}$$

Sillen (1961) estimated the buffering capacity of these silicates to be about 1 mole per liter or approximately 2000 times the buffering capacity of carbonates.

Alkalinity is the sum of all alkaline components minus the sum of all acidic components. In aquatic chemistry, we are particularly interested in the amount of hydrogen ions that we have to add to a considered aquatic solution to obtain a pH corresponding to an aquatic solution of carbon dioxide. This is the alkalinity defined in Figure 7.6.

Alkalinity of many natural aquatic systems is often (but not always) with good approximation equal to the hydrogen carbonate concentration because hydrogen carbonate is at a pH between 5 and 8.5, the dominating component yielding alkalinity in many natural waters. Other ions such as chloride and sulfate do not contribute to the alkalinity.

Note that we cannot find the pH for a mixture of two solutions with known pH, as the weighted average of the two pH values, because the resulting number of free hydrogen ions depends on the composition of the two solutions: which ions would be able to react with the free hydrogen and hydroxide ions? We can therefore only find the pH for a mixture when we know the concentrations of alkaline and acidic components in the two solutions and can calculate the possible neutralization reactions. The alkalinity can, however, be found easily for a mixture of two solutions because alkalinity is based on a "book keeping" of *all* alkaline and acidic components, that is, all components that can participate in acid–base reactions in the actual pH range. If, for instance, the alkalinity is 4 meq/L for one solution and 8 meq/L for another solution and we mix equal volumes of the two solutions, the resulting alkalinity will be 6 meq/L, which can be used to find the resulting pH for the mixture.

7.11 Dissolved Carbon Dioxide

Open aquatic systems have, in addition to the acid–base reactions of the dissolved components, an equilibrium between carbon dioxide in the atmosphere and dissolved in water. Henry's equation (7.28) can be applied to find the concentration of carbon dioxide in water. This equation implies that the carbon dioxide concentration is constant, independent of pH. Part of the dissolved carbon dioxide (in the order of 1%) reacts with water and forms carbon acid. It is, however, in most calculations convenient not to distinguish the dissolved carbon dioxide and the carbon acid but to consider the total amount of carbon dioxide and carbon acid. The usually applied pK values are based on this assumption. Henry's constant and the pK values for carbon acid and hydrogen carbonate are dependent on the temperature (see Table 7.4).

Figure 7.7 shows a double logarithmic diagram for an open aquatic system in equilibrium with the carbon dioxide in the atmosphere. The figure

TABLE 7.4

Equilibrium Constants of Various Carbon Dioxide–Carbonate Equilibria at I = 0

Type of Constant	5°C	10°C	15°C	20°C	25°C	40°C
Solubility product of $CaCO_3$	8.35	8.36	8.37	8.39	8.42	8.53
pK_1 for H_2CO_3	6.52	6.46	6.42	6.38	6.35	6.35
pK_2 for H_2CO_3	10.56	10.49	10.43	10.38	10.33	10.22
pK_w	14.73	14.53	14.34	14.16	14.00	13.53
K_H	1.20	1.27	1.34	1.41	1.47	1.64

illustrates the system for the years 1999–2002, where the partial pressure of carbon dioxide was about 0.000385 atm corresponding to a concentration of 385 ppm on volume/volume basis. During 2013/2014, the concentration was approximately 400 ppm. Room temperature is presumed in the diagram.

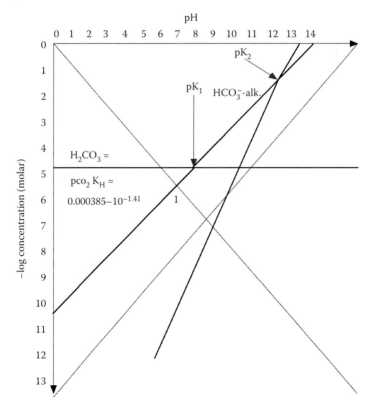

FIGURE 7.7
A double logarithmic diagram for an open aquatic system in equilibrium with the carbon dioxide in the atmosphere. A concentration of 385 ppm, corresponding to the carbon dioxide concentration for the year 1999/2000, is presumed.

Total carbonic acid = carbonic acid + dissolved carbon dioxide, denoted by C_T, will therefore be in a double logarithmic diagram corresponding to a horizontal line at $\log C = 0.000385 \cdot 10^{-1.41}$. The line representing hydrogen carbonate has a slope of +1 and intersects the carbon acid line at $pH = pK_1$. The carbonate line will correspondingly have a slope of +2 and intersect the hydrogen carbonate line at $pH = pK_2$.

The composition of an open aquatic system can be easily found by the application of Figure 7.7, provided that the alkalinity is known and other alkaline components can be omitted. If, for instance, the alkalinity is found to be 0.001 M and it is assumed that pH is below 9, the alkalinity will with good approximation be equal to the hydrogen carbonate concentration. It is seen in Figure 7.7 that a hydrogen carbonate concentration of 0.001 corresponds to a pH of 8.0—the assumption that pH is below 9.0 is therefore correct. The concentrations of carbon acid and carbonate can be easily read from the figure in this case.

EXAMPLE 7.4

Construct a double logarithmic diagram for carbonic acid–hydrogen carbonate–carbonate in a freshwater system at 10°C in equilibrium with the atmosphere containing 450 ppm carbon dioxide.

Mark on the diagram a stream with alkalinity 0.0025 eqv/L and indicate the corresponding pH.
Mark on the diagram a stream with alkalinity 0.0005 eqv/L and indicate the corresponding pH value.
What is the resulting pH in a lake receiving a mixture of equal volumes of the two streams?

Solution

At moderate pH, the alkalinity = $[HCO_3^-]$. In the diagram (see Figure 7.8), $[HCO_3^-] = 0.0025$ M (point 1) corresponding to pH = 8.5 is shown. Point 2 indicates $[HCO_3^-] = 0.0005$ M corresponding to pH = 7.8.

The alkalinity is conserved when two streams are mixed. Therefore, the alkalinity in the lake $\approx [HCO_3^-] = 0.0015$ M at pH = 8.3 (point 3).

In addition to the equilibrium between carbon dioxide in the atmosphere and carbon acid in water, solid carbonate may be presented as suspended matter and/or in the sediment and be in equilibrium with the calcium and carbonate ions in water according to the solubility product:

$$[Ca^{2+}]\left[CO_3^{2-}\right] = K_S = 10^{-8.4} \tag{7.43}$$

It is possible to also include this equation in the double logarithmic diagram as shown in Figure 7.9. The calcium ion line gets a slope of +2 and intersects the carbonate line at $\log C = -4.2$. The composition under these circumstances is found by a charge balance. The sum of the negative

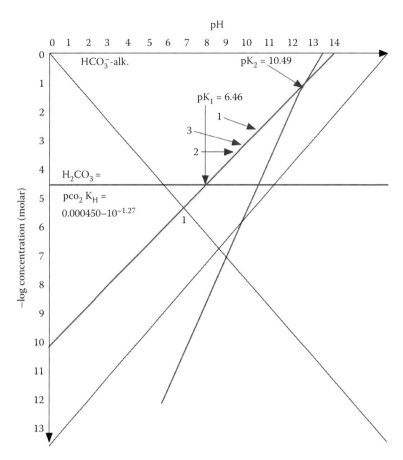

FIGURE 7.8
The solution to Example 7.4.

and positive ions must balance. In most aquatic systems, at pH < 9.0, the dominant cations will be the calcium ions and the dominant anions will be the hydrogen carbonate ions. This implies that the following equation with good approximation is valid:

$$2[Ca^{2+}] = \left[HCO_3^- \right] \qquad (7.44)$$

This corresponds in Figure 7.10 to a pH = 8.4. The carbonate ion concentration is negligible at this pH value. Note that in this case where two equilibria are imposed on the system (equilibrium with carbon dioxide in the atmosphere and equilibrium with solid calcium carbonate), the composition of water is given and no information of alkalinity is needed. The concentration of the carbonate ions must be included in the charge balance at higher pH, which is possible by an iteration.

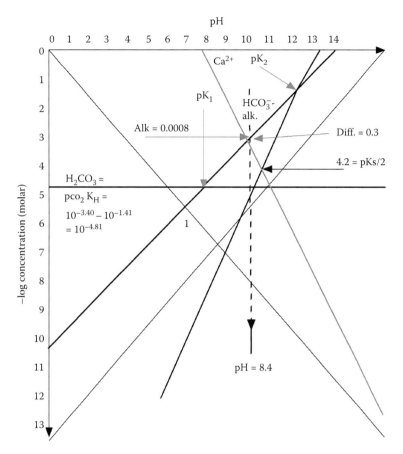

FIGURE 7.9
A double diagram for an aquatic system simultaneously in equilibrium with carbon dioxide in the atmosphere and solid calcium carbonate.

EXAMPLE 7.5

What is the composition when an open freshwater system is in equilibrium with solid calcium carbonate and carbon dioxide at $10^{-3.41}$ atm?

Solution

Figure 7.9 can be applied.
    The process determining the dissolution of calcium carbonate:

$$CaCO_3(s) + CO_2 + H_2O \rightarrow Ca^{2+} + 2HCO_3^-$$

The charge balance:

$$2[Ca^{2+}] + [H^+] = \left[HCO_3^-\right] + 2\left[CO_3^{2-}\right] + [OH^-]$$

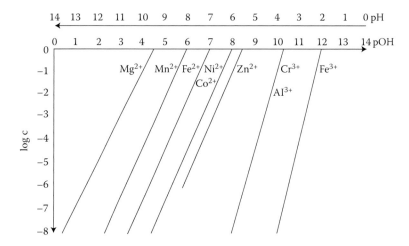

FIGURE 7.10
The solubility of several heavy metal cations as a function of pH.

With good approximation:

$$2[Ca^{2+}] = \left[HCO_{3^-}\right]$$

$$pH \approx 8.4; \left[HCO_{3^-}\right] \approx 10^{-3}\,M;\, [H_2CO_2{}^*]$$

$$\approx 10^{-5}\,M;\, [Ca^{2+}] \approx 5 \cdot 10^{-4}\,M;\, \text{and}\, \left[CO_3^{2-}\right] \approx 10^{-5}\,M$$

7.12 Precipitation and Dissolution: Solubility of Hydroxides

The solubility products of hydroxides, oxides, and carbonates have a particular interest in aquatic chemistry because these anions are present in high concentrations in natural aquatic systems. The hydroxides are furthermore of interest because they are insoluble in most of the heavy metals, and are utilized in the treatment of industrial wastewater containing heavy metals. As it has been shown in Figure 7.10, it is also possible to apply the double logarithmic plot to represent a solubility product. This is also illustrated in Figure 7.10, where the solubility of several heavy metal ions as a function of pH is due to the precipitation of hydroxides. The solubility products for several heavy metal hydroxides are shown in Table 7.5.

In many cases, the cations are able to form hydroxo complexes with hydroxide ions and thereby change the solubility that is found directly from the solubility product. The actual reactions for aluminum ions are

TABLE 7.5

pKs Values at Room Temperature for Metal
Hydroxides pKs = –log Ks, where Ks = [Mez+][OH–]z

Hydroxide	z = Charge of Metal Ion	pKs
$AgOH(1/2Ag_2O)$	1	7.7
$Cu(OH)_2$	2	20
$Zn(OH)_2$	2	17
$Ni(OH)_2$	2	15
$Co(OH)_2$	2	15
$Fe(OH)_2$	2	15
$Mn(OH)_2$	2	13
$Cd(OH)_2$	2	14
$Mg(OH)_2$	2	11
$Ca(OH)_2$	2	5.4
$Al(OH)_3$	3	32
$Cr(OH)_3$	3	32

**HOW TO MAKE A DOUBLE LOGARITMIC DIAGRAM FOR THE SOLUBILITY
AS INFLUENCED BY HYDROLYSIS/FORMATION OF HYDROXO COMPLEXES**

1. Set up a diagram indicating the initial components and how they can react to
   form various species. Example:

Species	$Al(OH)_3$	#H+	Log K*
Al^{3+}	1	3	8.5
$Al(OH)^{2+}$	1	2	3.53
$Al(OH)_2^+$	1	1	−0.8
$Al(OH)_3$	1	0	−6.5
$Al(OH)_4^-$	1	−1	−14.5

2. Find the equilibrium constants (log $*K_1$, log $*K_2$, log $*K_3$, log $*K_4$) for the
   reactions between the aluminum hydroxide and one or more hydrogen ions,
   here exemplified by $Al(OH)_3 + 3H^+ \leftrightarrow Al^{3+} + 3H_2O$.

   $*K_1$ can be found from the solubility product, $K_s = [Al^{3+}][OH^-]^3$:

   $$K_s = \frac{[Al^{3+}][OH-]^3[H+]^3}{[H^+]^3} = K_1 \cdot K_w^3$$

   where K_w is the ion product of water = 10^{-14} at room temperature.

3. Plot in a double logarithmic diagram log (species) versus pH by use of the
   equilibrium constants for the formation of the considered species. The
   solubility of Al(III) is the sum of all the soluble species, Al^{3+}, $Al(OH)^{2+}$, etc.

FIGURE 7.11
How to utilize a double logarithmic diagram when hydroxo complexes are formed.

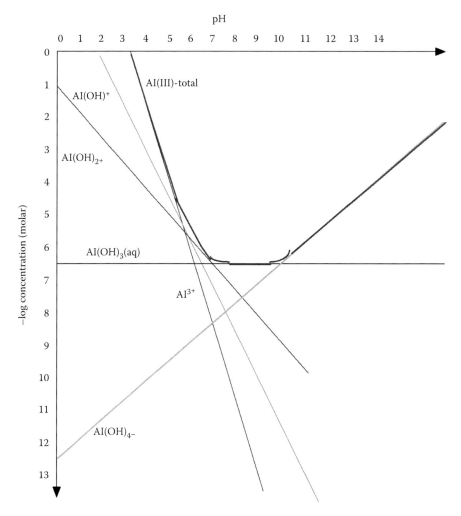

FIGURE 7.12
The solubility of aluminum(III) considering the formation of several hydroxo complexes as a function of pH.

shown in Figure 7.11. The figure illustrates how to cope with the solubility as a function of pH if hydroxo complexes can be formed. The double logarithmic diagram corresponding to the presented procedure is shown in Figure 7.12.

The solubility of iron(III) hydroxide, zinc oxide, and copper oxide can be found by the same method. The diagram for iron(III) hydroxide is shown in Figure 7.13.

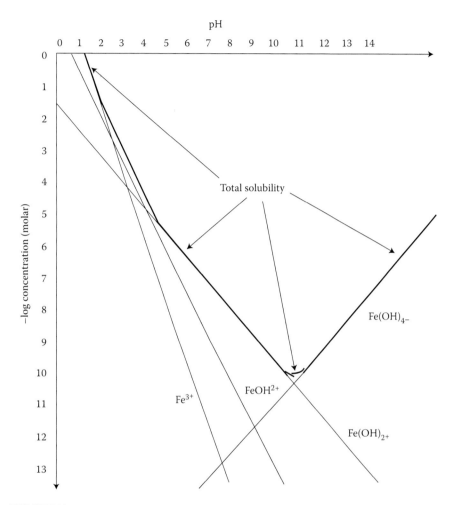

FIGURE 7.13

The solubility of amorphous iron(III) hydroxide as a function of pH. The construction of the diagram is as explained for aluminum(III) hydroxide. The solubility of iron(III) hydroxide is determined by the formation of iron(III) ions and formation of complexes with one, two, or four hydroxide ions.

7.13 Solubility of Carbonates in Open Systems

The solubility of calcium carbonate in open aquatic systems has already been dealt with in Section 7.10. Figure 7.14 illustrates strontium carbonate, iron(II) carbonate, cadmium carbonate, and zinc carbonate, similar to Figure 7.9 for calcium carbonate.

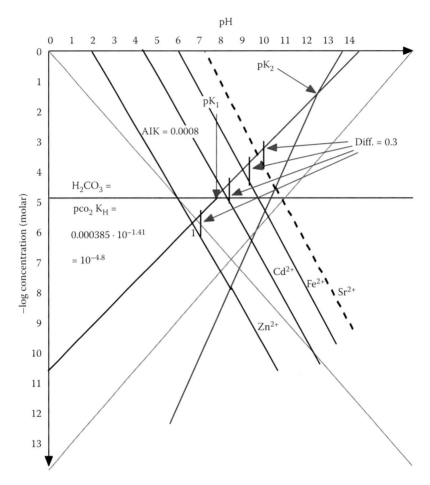

FIGURE 7.14
Double logarithmic representation of an equilibrium of solid strontium carbonate, iron(II) carbonate, cadmium carbonate, and zinc carbonate in an open aquatic system. The composition will correspond to the lines where the difference between the concentration of the metal ion and hydrogen carbonate is 0.3, corresponding to the equation $2[Me^{2+}] = [HCO_3^-]$.

7.14 Solubility of Complexes

Complex formation will be covered later in this chapter, but here it will be shown how a simple complex formation under simultaneous reaction with a compound with very little solubility can enhance the solubility considerably. The calculations are made by adding two suitable reactions and determining the equilibrium constant for the total process as the product of the equilibrium constants of the two added processes. If the equilibrium constant for the total process is high, the solubility will with very good approximation be very pronounced. If the equilibrium constant is not very high, the solubility is made the unknown in a suitable equation.

The method is shown by the below example.

EXAMPLE 7.6

Find the solubility of iron(III) in a freshwater lake with and without the presence of 10 µmol citrate/L. The ion strength of the lake water is 0 and pH = 9.0. The solubility product of iron(III) hydroxide is 10^{-24} mol^2/L^2. The formation constant for the complex between citrate and iron(III) at pH = 9.0, where citric acid has dissociated all hydrogen ions with good approximation, is 10^{28}.

Solution

From the solubility product, it is found that the solubility at pH = 9.0 (the hydroxide concentration is 10^{-5} M) is 10^{-9} M. The solubility of iron(III), when citrate is present, is determined by the reaction

$$Fe(OH)_3 + citrate^{3-} <-> Fe(III)\, citrate^- + 3OH^-$$

and the mass equation yields

$$\frac{[Fecitrate^-][OH^-]^3}{[citrate^{3-}]} = 10^{+4} = \frac{X * 10^{-15}}{(0.00001 - X)}$$

$$\frac{X}{(0.00001 - X)} = 10^{+19}$$

X must be 10 µM/L with good approximation, which of course also corresponds to the concentrations of citrate as it is the equilibrium constant (10^{+19}) that determines how much of iron(III) will dissolve.

7.15 Stability of the Solid Phase

It is not possible by a comparison of the numerical values of the equilibrium constant to decide which of the two or more solid phases control the solubility of an ion. It is necessary to determine by calculation which solid phase gives the smallest concentration of the considered ion and it will be the solid phase with the highest stability. Example 7.7 illustrates a case study: Is it iron(II) carbonate or iron(II) hydroxide that determines the solubility of iron(II) in natural water containing carbonate?

EXAMPLE 7.7

Is it iron(II) carbonate or iron(II) hydroxide that determines the solubility of iron(II) in natural anoxic water at pH = 6.8 with an alkalinity of 10^{-4}? The solubility products are $10^{-10.7}$ mol^2/L^{-2} and $10^{-14.7}$ mol^3/L^3, respectively. The pK for hydrogen carbonate is 10.1.

Solution

The iron(II) concentration resulting from the solubility of iron(II) carbonate and iron(II) hydroxide is determined as

$$FeCO_3(s) = Fe^{2+} + CO_3^{2-} \qquad \log K_s = -10.4$$

$$H^+ + CO_3^{2-} = HCO_3^- \qquad -\log K_2 = 10.1$$

For the process

$$FeCO_3 + H^+ = Fe^{2+} + HCO_3^-$$

the equilibrium constant is found by adding the processes, which implies that $\log {}^*K_s = -0.3$. It means that $\log [Fe^{2+}] = \log {}^*K_s - pH - \log [HCO_3^-] = -0.3 - 6.8 - (-4) = -3.1$.

$$Fe(OH)_2(s) = Fe^{2+} + 2OH^- \qquad \log K_s = -14.5$$

$$2H^+ + 2OH^- = 2H_2O \qquad -2\log K_w = +28.0$$

For the process

$$Fe(OH)_2(s) + 2H^+ = Fe^{2+} + 2H_2O$$

the equilibrium constant is found by adding the two processes, which implies that $\log {}^*K_s = 13.5$ and $\log [Fe^{2+}] = \log {}^*K_s - 2pH = 0.1$.

The calculations show that iron(II) will be present under the given condition in the concentration $10^{-3.1}$, determined by a solid phase consisting of iron(II) carbonate, with the mineralogical name siderite.

It is possible under the given conditions to find the pH value at which iron(II) hydroxide will become the most stable solid and thereby determine the iron(II) solubility. The two expressions for the iron(II) concentration yield $-0.3 - pH - (-4) = 13.5 - 2pH$. At pH = $13.5 + 0.3 - 4 = 9.8$, the two solid phases will have the same stability, but above this pH value, iron(II) hydroxide will determine the iron(II) solubility. At pH = 9.8, the solubility of iron(II) is $10^{-6.1}$.

7.16 Complex Formation

Any combination of cations with molecules or anions containing free pairs of electrons is named coordination formation. The metal ion is denoted as the central atom and the anions or the molecules are called the ligands. The atom responsible for the coordination is called the ligand atom. If a ligand contains more than one ligand atom and thereby occupies more than one coordination position in the complex, it is referred to as multidentate. Ligands occupying one, two, three, four, and so on are named unidentate, bidentate, tridentate, tetradentate, and so on. Complex formation with multidentate ligands is called chelation and the complexes are called chelates. Typical examples from aquatic chemistry are

Oxalate and ethylenediamine (bidentate)

Citrate (tridentate)

Ethylenediamine tetraacetate (hexadentate)

If there is more than metal ion in the complex, we are talking about polynuclear complexes.

Complex formation takes place according to the following reaction scheme:

$$Me^{n+} + L^{m-} \rightarrow MeL^{(n-m)+} \tag{7.45}$$

where Me is a metal and L is a ligand. The mass equation gives the following expression:

$$\frac{[MeL^{(n-m)+}]}{[Me^{n+}][L^{m-}]} = K \tag{7.46}$$

K is named the stability constant, complexity constant, or formation constant.

The coordination number for a metal is the number of coordination position on the metal—positions where the free electron pairs can attach the ligand to the metal. Even coordination numbers are most frequently found. Coordination numbers for the metal ions most frequently found in aquatic environment are listed in Table 7.6.

7.17 Environmental Importance of Complex Formation

Complex formation has a great influence on the effects of metal ions on aquatic organisms and ecosystems. Complex formation has the following effects:

1. The solubility can be increased:

$$MeY(S) + L = MeL + Y \tag{7.47}$$

TABLE 7.6

Coordination Numbers for Metals of
Interest in Aquatic Chemistry

Metal	Coordination Numbers
Cu^+	2
Ag^+	2
Hg_2^{2+}	2, 4
Li^+	4
Be^{2+}	4, 6
Al^{3+}	4, 6
Fe^{3+}	4, 6
Cu^{2+}	4, 6
Ni^{2+}	4, 6
Hg^{2+}	2, 4
Fe^{2+}	6
Mn^{2+}	4, 6

Me represents a metal, Y a sediment, and L a ligand that is able to react with the metal ion.

2. The oxidation stage of the metal ion may be changed. The mass equation constants for the following two processes may differ:

$$Me^{n+} + me^- \rightarrow Me^{(n-m)+} \tag{7.48}$$

$$MeL^{n+} + me^- \rightarrow MeL^{(n-m)+} \tag{7.49}$$

if the stability constant (mass equation) for the two complexes in Equation 7.49 are different.

3. The metal toxicity may be changed because the complexes have another bioavailability than the metal ions.

4. The ion exchange and adsorption processes of the metal ions may be changed. The adsorption isotherms and ion exchange equilibrium constant will in far most cases be different for the complexes and the metal ions due to the bigger size of the complexes among other properties.

The triangle diagram in Figure 7.15 shows which forms of metal ions are usually found in freshwater and saltwater. The difference is due to the high salinity (chloride concentration) and ion strength in saltwater and the presence of organic ligands and a higher concentration generally of suspended matter in freshwater.

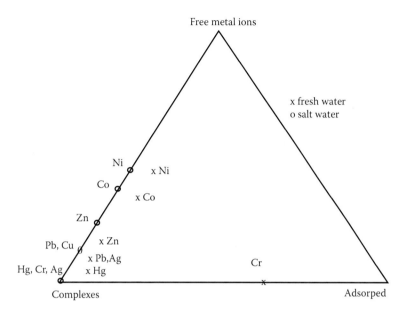

FIGURE 7.15
Triangle diagram for the form of heavy metal ions in freshwater and saltwater. Presence of ligands in aquatic systems will generally enhance the transfer of metal ions from sediment, soil, and suspended matter to water, but the toxicity will generally be reduced.

7.18 Conditional Constant

The complex formation in aquatic systems is often very complicated because a number of side reactions, including acid–base reactions, precipitations, and redox reactions, are possible in addition to the main reaction, the complex formation. Reactions between metal ions and ligands often determine the release of metal ions from sediment, pH often determines the strength of a complex, and the bioavailability of toxic substances is highly dependent on the form of the compound—complex bound or free, base or acid form, and so on. It is therefore of importance to be able to find the concentrations of the various forms in aquatic systems under given conditions (Alloway and Ayres, 1993).

It is possible to consider the side reactions by the application of what is called conditional constants. Let us consider the reaction

$$Me^{n+} + L^{m-} \leftrightarrow MeL^{(n-m)+} \tag{7.50}$$

It is assumed that the following side reactions are possible:
A precipitation:

$$pMe^{n+} + nY^{p-} \leftrightarrow Me_pY_n(s) \tag{7.51}$$

An acid–base reaction:

$$L^{m-} + H^+ \leftrightarrow HL^{(m-1)-}; \quad HL^{(m-1)} + H^+ \leftrightarrow H_2L^{(m-2)-} \tag{7.52}$$

f_{Me}, defined as the ratio of the metal ion concentration to the metal concentration, including other forms determined by the side reactions, can be found from the solubility product, K_S, if $[Y^{P-}]$ e is known. If concentrations are used, we get

$$[Y^{P-}]^n [Me^{n+}]^P = K_S \tag{7.53}$$

$$\frac{[Me^{n+}]}{[Me']} = f_{Me} \frac{K_S^{1/P}}{[Y^{P-}]^{n/P}} * \frac{1}{[Me']} \tag{7.54}$$

where the symbol Me′ represents all the metal forms except the complex, formed by what is considered the main reaction (7.50).

f_L, defined as the ratio between L^{m-} and the ligand concentration, including other forms determined by side reactions, can be found in a similar manner:

$$\frac{[L']}{[L^{m-}]} = \frac{1}{f_L} = 1 + \frac{[H^+]^m}{K_1^* K_2 \dots K_m} + \frac{[H^+]^{m-1}}{K_2^* K_3 \dots K_m}$$
$$+ \frac{[H^+]^{m-2}}{K_3^* K_4 \dots K_m} + \dots + \frac{[H^+]}{K_m} \tag{7.55}$$

where L′ symbolizes all ligand forms except the complex formed by the main reaction (7.16). K_1 to K_m are the dissociation constants for the stepwise dissociation of hydrogen ions for the acid $H_m L$.

A reformulation takes place:

$$\frac{[MeL^{(n-m)+}]}{[Me^{n+}][L^{m-}]} = K = \frac{[MeL^{(n-m)+}]}{[Me']* f_{Me}^* [L']* f_L}$$
$$\frac{[MeL^{(n-m)+}]}{[Me'][L']} = K^* f_{Me} f_L = K_{cond} \tag{7.56}$$

K_{cond} is denoted the conditional equilibrium constant. If K_{cond} is known, it is possible to find the concentrations of free metal ions and metal complexes by the use of Equations 7.55 and 7.56. K_{cond} is of course only valid under the conditions given by pH and $[Y^{P-}]$. However, if these two concentrations are known—the conditions are known—then the usual mass constant calculations can be carried out using K_{cond} instead of K.

The most important equations for calculations of complex formation are summarized in Figure 7.16. Note that K is used for the mass equation constant where one ligand is attached to the metal ion, whereas β_n is used for the reaction between the metal ion and n ligand.

> ### COMPLEXITY CONSTANT, STABILITY CONSTANTS, AND FORMATION CONSTANT:
>
> $K_1 = [ML]/[M][L]$; $K_2 = [ML_2]/[ML][L]$; $K_3 = [ML_3]/[ML_2][L]$, etc.
>
> $\beta_1 = K_1$; $\beta_2 = K_1 K_2$; $\beta_3 = K_1 K_2 K_3$, etc.
>
> K^* and β^* symbolize the constants where HL is the reactant (a ligand with only one H is considered in this example).
> If f_L is the fraction of the L-form at a given pH of the total concentration, that is, $L + HL = L'$, the conditional constants are
>
> $K_{1cond} = [ML]/[M][L'] = f_L K_1$
> $K_{2cond} = [ML_2]/[ML][L'] = f_L K_2$
> $K_{3cond} = [ML_3]/[ML_2][L'] = f_L K_2$, etc.
>
> Side reactions by M are accounted for in a similar manner.

FIGURE 7.16
Important equations by equilibrium calculations of complex formations.

EXAMPLE 7.8

The total concentration of mercury in a lake at $pH = 6.5$ is determined to be 0.05 mM. The equilibrium constants for formation of $Hg(OH)^+$ and $Hg(OH)_2^0$ by hydrolysis are $10^{-3.7}$ M and $10^{-2.6}$ M. $M_{Hg} = 200.59$ g/mol.

  A. At which concentrations can the various forms of mercury(II) be found under these circumstances?
     The LC_{50} value for Hg^{2+} ions is 1.2 mg/L for daphnia.
  B. Is the LC_{50} value exceeded, when it is assumed that the hydroxo complexes are 100 times less toxic than Hg^{2+} ions?

Solution

  A. Two reactions forming hydroxo complexes:

$$Hg^{2+} + H_2O \rightarrow Hg(OH)^+ + H^+; \quad K_1 = 10^{-3.7} \text{ M, } pK_{s1} = 3.7$$

$$Hg(OH)^+ + H_2O \rightarrow Hg(OH)_2^0 + H^+; \quad K_2 = 10^{-2.6} \text{ M, } pK_{s2} = 2.6$$

At $pH = 6.5$, $Hg(OH)_2^0$ is the dominating form of mercury(II) according to the two equilibrium constants, that is, $[Hg(OH)_2^0] \approx$ 0.05 mM.

The concentrations of $[Hg(OH)^+]$ and $[Hg^{2+}]$ are found from $[Hg(OH)_2^0] \approx 0.05$ mM and the equilibrium constants

$$[Hg(OH)^+] = \frac{\left[Hg(OH)_2^0\right] \cdot 10^{-pH}}{K_2} = 10^{-8.2} \, M$$

$$[Hg^{2+}] = \frac{[Hg(OH)^+] \cdot 10^{-pH}}{K_1} = 10^{-11} \, M$$

B. The hydroxo complexes are 100 times less toxic, that is, they have an LC_{50} value of 0.12 g/L = 0.55 mM.
The ratio between concentrations and LC_{50} values is found:

$$\frac{(10^{-8.2} \, M + 0.05 \, mM)}{0.55 \, mM} + \frac{10^{-11} \, M}{0.0055 \, mM} = 0.09 < 1.2$$

The LC_{50} for daphnia is not exceeded.

7.19 Application of Double Logarithmic Diagrams to Determine the Conditional Constants for Complex Formation

Let us consider a complex formation according to the following reaction scheme:

$$Me^{n+} + L^{m-} = MeL^{(n-m)+} \quad \log K = F$$

exemplified by (glycinate = Gly):

$$Fe^{3+} + Gly^- = FeGly^{2+} \quad \log K = 10.8$$

The conditional constant, $K_{cond} = [FeGly^{2+}]/[Fe'] \, [Gly']$.
The following equations can be applied to find the conditional constant as a function of pH:

$$[Fe'] = [Fe(III)_{total}] - [FeGly^{2+}] \tag{7.57}$$

$$[Fe'] = [Fe^{3+}] + [FeOH^{2+}] + \left[Fe(OH)_2^+\right] + [Fe(OH)_3] + \left[Fe(OH)_4^-\right] \tag{7.58}$$

$$[Fe'] = [Fe^{3+}]\left(1 + \frac{^*K_1}{[H^+]} + \frac{^*\beta_2}{[H^+]^2} + \frac{^*\beta_3}{[H^+]^3} + \frac{^*\beta_4}{[H^+]^4}\right) \tag{7.59}$$

$$[Fe'] = \frac{[Fe^{3+}]}{f_{Fe}} \tag{7.60}$$

$$[Gly'] = [Gly_{total}] - [FeGly^{2+}] \tag{7.61}$$

$$[Gly'] = [H_2Gly^+] + [Hgly] + [Gly^-] \tag{7.62}$$

$$[Gly'] = \frac{[Gly^-](1+[H^+])}{K_2} + \frac{[H^+]^2}{K_1K_2} \tag{7.63}$$

$$[Gly'] = \frac{[Gly^-]}{f_{gly}} \tag{7.64}$$

Based on these equations, it is possible to calculate f_{Fe} and f_{Gly} as a function of pH, provided of course that the various applied mass equations constants are known: $\log {}^*K_1 = -3.05$, $\log {}^*\beta_2 = -6.31$, $\log {}^*\beta_3 = -13.8$, and $\log {}^*\beta_4 = -22.7$. The pK values for H_2Gly^+ are 3.1 and 9.9.

Figure 7.17 shows f_{Fe}, f_{Gly} and the product of the two, $f_{Fe} f_{Gly}$, calculated as a function of pH. Equations 7.59, 7.60 and 7.63, 7.64 are applied. f_{Fe}, f_{Gly}, and

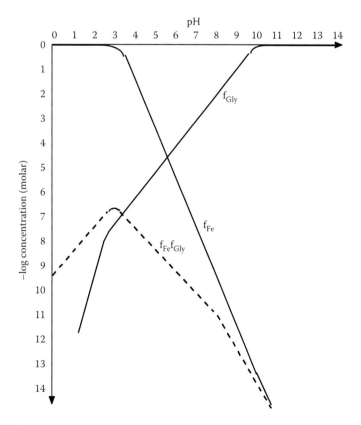

FIGURE 7.17
f_{Fe} and f_{Gly} and the product $f_{Fe}f_{Gly}$ found as a function of pH by Equations 7.23 through 7.30.

the product $f_{Fe} f_{Gly}$ are found at a given pH and the conditional constant is calculated as $K_{cond} = K f_{Fe} f_{Gly}$. K_{cond} has a maximum around pH = 3.2. At this pH, $\log K_{cond} = \log K + \log f_{Fe} f_{Gly} = 10.8 - 7.0 = 3.8$.

EXAMPLE 7.9

A. By applying the diagrams in Figure 7.18, answer the following questions:

What is the dominant form of cadmium(II) in water with different salinities? $M_{NaCl} = 58.44$ g/mol.

What is the LC_{20} value for blue mussels for total cadmium in mg/L when the following information is available: LC_{20} for Cd^{2+} ions is 2.5 mg Cd/L, 25 mg Cd/L for the monochloro complex, 60 mg Cd/L for the dichloro complex, 75 mg Cd/L for the trichloro complex, and 120 mg Cd/L for the tetrachloro complex? It is presumed that no synergistic or antagonistic effect will occur. The salinity can with good approximation be referred to sodium chloride and the specific density of seawater is independent of the salinity 1.0 kg/L.

Use the following table to answer the questions:

Salinity	5‰ (Baltic Sea 200 km North of Stockholm)	12‰ (Baltic Sea at the Møn Island)	20‰ (Kattegat by Anholt)	40‰ (Mediterranean Sea)
Dominating form of Cd(II)				
Approximate LC_{20} value in mg/L				

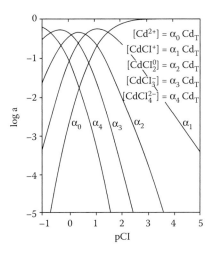 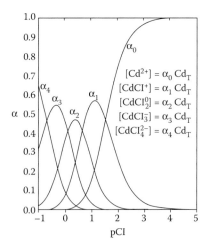

FIGURE 7.18
Cadmium chloride complexes as a function of the chloride concentration; see Example 7.9. pCl is −log (chloride concentration).

Solution

A. Use the diagrams to answer the questions. The answer can be found in the table.

B. For the calculation of $LC_{20,Cd(II),pCl}$ for the four pCl values read the α values. Calculate $LC_{20,Cd(II),pCl}$ as

$$\frac{1}{LC_{20,Cd(II),pCl}} = \sum_{n=0}^{n=4} \frac{\alpha_n}{LC_{20,Cd[Cl]_n^{2-n}}}$$

Salinity	5‰ (Baltic Sea 200 km North of Stockholm)	12‰ (Baltic Sea at the Møn Island)	20‰ (Kattegat by Anholt)	40‰ (Mediterranean Sea)
pCl	1.07	0.69	0.47	0.16
Dominating form of Cd(II)	$CdCl^+$	$CdCl^+$ and $CdCl_2^0$	$CdCl_2^0$	$CdCl_2^0$ and $CdCl_3^-$
Approximate LC_{20} value in mg/L	9.4 mg/L	17.6 mg/L	30.4 mg/L	45.9 mg/L

7.20 Redox Equilibria: Electron Activity and Nernst Law

There is an analogy between acid–base and reduction–oxidation reactions. Acid–base reactions exchange protons. Acids are proton donors and bases are proton acceptors. Redox reactions exchange electrons. Reductants are electron donors and oxidants are electron acceptors. As there are no free hydrogen ions, protons, there are no free electrons. This implies that every oxidation is accompanied by a reduction and vice versa. Just as pH has been introduced as the proton activity, we may introduce an electron activity defined as

$$pe = -\log\{e^-\} \tag{7.65}$$

where e^- is the electron and p as usual is an abbreviation of $-\log$. As free electrons do not exist, pe should be considered a concept that is introduced to set up a parallel to pH and facilitate the calculations of redox equilibria. The relationship between pe and the redox potential, E, is

$$pe = \frac{F * E}{2.3RT} \tag{7.66}$$

where F is Faraday's constant $= 96485$ C/mol $=$ the charge of 1 mol of electrons, and R is the gas constant $= 8.314$ J/mol K $= 0.082057$ l atm/mol K. Nernst law

$$E = E° + \frac{RT}{nF} \log \frac{\{ox\}}{\{red\}} \tag{7.67}$$

may be rewritten by the use of $pe°$:

$$pe° = \frac{F * E°}{2.3RT} \tag{7.68}$$

to

$$pe = pe° + \left(\frac{1}{n}\right) \log \frac{\{ox\}}{\{red\}} \tag{7.69}$$

This yields furthermore the following relationship between the free energy and pe, as $\Delta G = -E\,n$:

$$pe = \frac{-\Delta G}{n * 2.3RT} \quad pe° = \frac{-\Delta G°}{n * 2.3RT} \tag{7.70}$$

Note that these equations are applied on the half reaction—that an oxidant takes up an electron, which is transferred to a reductant. A redox process, however, requires as pointed out above, that another reductant is available to deliver the electron and thereby that this reductant is changed to the corresponding oxidant.

Equation 7.69 implies that the following relationship between pe and the equilibrium constant, K, for the half reaction, in which the electron takes part, is valid:

$$pe° = \frac{1}{n} \log K \tag{7.71}$$

As an illustration of the application of these equations, the following reaction is considered:

$$Fe^{3+} + e^- = Fe^{2+} \tag{7.72}$$

The standard redox potential $E°$ corresponding to $\{ox\} = \{red\} = 1$ for this process can be found in any table of standard redox potentials as 0.77 V at room temperature. It means

$$E° = 0.77; \ pe° \frac{F * 0.77}{2.3 * R * 298} = (t = 25°C) \tag{7.73}$$

$$\log K = n * pe° = 1 * \frac{0.77}{0.059} = 13.0 \tag{7.74}$$

where $\log K = \log (\{Fe^{2+}\}/\{Fe^{3+}\}\{e^-\})$.

If, for instance, in acidic solutions, $[Fe^{3+}] = 10^{-3}$ and $[Fe^{2+}] = 10^{-2}$, applying concentrations instead of activities as a reasonably good approximation, we get

$$pe = pe^o + \left(\frac{1}{n}\right) \log \frac{\{ox\}}{\{red\}} = 13 + 1 * \log 0.1 = 12.0 \qquad (7.75)$$

Figure 7.19 summarizes the equations.

EXAMPLE 7.10

Find the pe value for a natural aquatic system at pH = 7.0 and in equilibrium with the atmosphere (partial pressure of oxygen = 0.21 atm). What is the ratio $\{Fe^{2+}\}$ to $\{Fe^{3+}\}$ in water?

Solution

The following process determines the redox potential:

$$0.5O_2 + 2H^+ + 2e^- = H_2O \qquad (7.76)$$

In tables, it is possible to find that log K for this process is 41.56 or the standard redox potential can be found and log K can be calculated based on E°.

Redox Equations:

$$E = E^\circ + 2.3\ RT/nF\ (\log\ \{ox\}/\{red\})$$

or

$$pe = pe^o + 1/n\ (\ \log\ \{ox\}/\{red\}\ )$$

$$\Delta G^\circ = -E^\circ\ nF = -RT\ \ln K =$$
$$2.3\ RT\ \log K = 2.3\ RT\ n\ pe^o$$

Note that when the composition, that is, $\{ox\}/\{red\}$ is given, the redox potential E or pe can be found.

Freshwater in equilibrium with the atmosphere (oxygen partial pressure 0.21 atm) will have the following pe:

$$pe = 20.78 + 0.5\ \log\ (0.21)^{0.5}\{H^+\}^2$$

From the pe value, the ratio between any redox pair in freshwater is determined.

FIGURE 7.19
Summary of the most important equations to calculate redox potential, pe, and the equilibrium constant.

It means that $pe° = 41.56/2 = 20.78$.

$$pe = pe° + 0.5 \log\left(\sqrt{p_{O_2}}\{H^+\}^2\right) = 20.78 + 0.5 \log\left(\sqrt{0.21 \cdot 10^{-14.0}}\right) =$$

$$20.78 + 0.5 (-14.34) = 13.61$$

This pe value determines all ratios between oxidants and reductants because the oxygen concentration in water in equilibrium with the atmosphere will inevitably have a constant concentration determined by the partial pressure of 0.21 atm. Therefore

$$1 [Fe^{3+}] 13.61 = pe = pe° + \frac{1}{n} \log \frac{[Fe^{3+}]}{[Fe^{2+}]} = 13.0 + \log 10^{0.61} \qquad (7.77)$$

The ratio $\{Fe^{2+}\}$ to $\{Fe^{3+}\} \approx 4.0$.

7.21 pe as Master Variable

Equation 7.68 can be used to set up a graphical presentation in a double logarithmic diagram of a redox process. How is the ratio of $\{Fe^{2+}\}$ to $\{Fe^{3+}\}$ changed with changing redox potential or pe? Obviously, at low pe, the reductant will dominate, and at high pe, the oxidant will dominate. When $pe = pe°$, the two forms will be equal. The double logarithmic diagram is therefore very similar to the double logarithmic diagram applied for acid–base reactions. A double logarithmic diagram for $\{Fe^{2+}\}$ and $\{Fe^{3+}\}$ is shown in Figure 7.20.

7.22 Examples of Relevant Processes in the Aquatic Environment

An aqueous solution at a given pH and a given pe determines the partial pressure of hydrogen and oxygen according to the following redox processes:

$$2H^+ + 2e^- = H_2(g)$$

$\log K = 0$ (the process is reference to the redox potential scale) $\qquad (7.78)$

$$2H_2O + 2e^- = H_2(g) + OH^- = \log K$$

$$= -28 \text{ (combine water's ion product } 10^{-14} \text{ with process (7.77)} \quad (7.79)$$

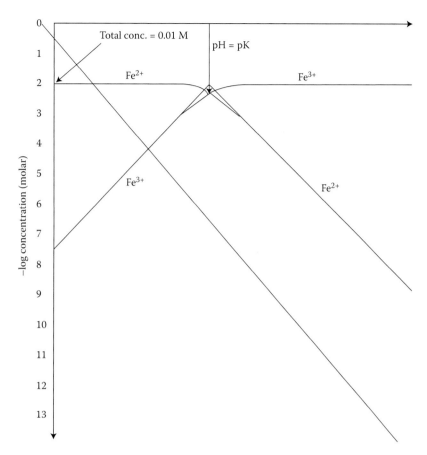

FIGURE 7.20
Double logarithmic diagram for the redox process: $Fe^{3+} + e^- = Fe^{2+}$.

$$O_2(g) + 4H^+ + 4e^- = 2H_2O \quad \log K = 83.1(4 \cdot 20.78) \tag{7.80}$$

$$O_2(g) + 4e^- + 2H_2O = 4OH^-$$

It implies that we obtain the following relationships between the partial pressure and pH and pe:

$$\log pH_2 = 0 - 2pH - 2 pe \tag{7.81}$$

$$\log pO_2 = -83.1 + 4pH + 4pe \tag{7.82}$$

If on the other side the partial pressure of oxygen can be considered constant and in equilibrium with aqueous solution and pH is given, the pe value can be found from Equation 7.81.

Manganese is present in aquatic systems as manganese dioxide (s) and manganese ions according to the following redox potential:

$$MnO_2(s) + 4H^+ + 2e^- \rightarrow Mn^{2+} + 2H_2O \qquad (7.83)$$

The concentration of manganese ions is determined by the redox potential for the aqueous solution. If the redox potential is determined by the partial pressure of oxygen in equilibrium with the aqueous solution, we obtain the following equation to determine the manganese ion concentration:

$$pe = 13.6 = 20.42 + 0.5 \cdot \log\left(\frac{[H+]^4}{[Mn^{2+}]}\right) \qquad (7.84)$$

$$\Downarrow$$

$$\log[Mn^{2+}] = 2 \cdot (20.42 - 13.6) - 7.4 = -14.4 \qquad (7.85)$$

Iron(II) is often under anaerobic conditions present in sediment and soil as iron(II) sulfide. If the sediment or soil is exposed to air, the following oxidation process takes place:

$$2FeS_2(s) + 2H_2O + 7O_2 = 2FeSO_4 + 2H_2SO_4 \qquad (7.86)$$

$$4FeSO_4 + O_2 + 2H_2SO_4 = 2Fe_2(SO_4)_3 + 2H_2O \qquad (7.87)$$

$$Fe_2(SO_4)_3 + 6H_2O = 2Fe(OH)_3(s) + 3H_2SO_4 \qquad (7.88)$$

These processes involve the formation of sulfuric acid and extreme low pH values may occur as a result. Simultaneously, the solid iron(II) sulfide is replaced by the solid iron(III) hydroxide.

The formation of iron(II) sulfide under anaerobic conditions may also mobilize phosphate stored in the sediment of aquatic systems. If sulfide is formed as a result of a low redox potential, the following process will take place:

$$FePO_4(s) + HS^- + e = FeS(s) + HPO_4^{2-} \qquad (7.89)$$

Chlorine is widely used as a disinfectant in aqueous solutions. Chlorine disproportionates (the same compound goes up and down in oxidation state):

$$Cl_2 + H_2O = HOCl + H^+ + Cl^- \qquad (7.90)$$

The effect of chlorine is associated with the formation of HOCl, which is a strong oxidant that is able to damage the enzyme system of bacteria. The hypochlorite ion, OCl^-, does not have this effect, which implies that pH is a determining factor for the disinfection effect of an aquatic chlorine solution. pK for HOCl at room temperature is 7.2. A pH above 8.0 is therefore not recommendable, when chlorine is applied as disinfectant.

7.23 Redox Conditions in Natural Waters

It is often convenient to consider the redox potential in natural aquatic systems, where it is presumed that pH = 7.0. A symbol pe^0 (w) is applied for these calculations. pe^0 (w) is analogous to pe^0, except that the activities of protons and hydroxide ions correspond to natural water at pH = 7.0. This implies that the following expression between pe^0 (w) and pe^0 is valid:

$$pe^0(w) = pe^0 + 0.5 \, n \log K_w \qquad (7.91)$$

where n is the number of moles of protons exchanged per mole of electrons.

$0.25 \, O_2(g) + H^+ + e^- = 0.5 \, H_2O$ has a pe^0 value of 13.75. The corresponding pe^0 (w) is therefore 20.75.

A table of pe^0 (w) values may be used to determine whether a system will tend to oxidize equimolar concentrations of any other system, which would require that it would have higher pe^0 (w). Sulfate/sulfide has a pe^0 (w) = −3.5, while CO_2/CH_2O has pe^0 (w) = −8.20. Sulfate is accordingly able to oxidize CH_2O in natural water.

EXAMPLE 7.11

The air in immediate contact with a wetland has a partial pressure of carbon dioxide and methane on 100 and 250 Pa, respectively. There is equilibrium between the water in the wetland and the air above the wetland. pH in the wetland is 7.0. The temperature is 25°C.

1. Find pe and the redox potential for the water in the wetland, when pe^0 (W) for the carbon dioxide/methane redox equilibrium is −4.13.
2. What is the ratio of iron(III) to iron(II) in water? pe^0 for Fe^{3+}/Fe^{2+} is 13.0.
3. How much will it change the redox potential (pe) if pH is changed from 7.0 to 7.6?
4. At a later stage, it is found that pH is 7.2 and the redox potential is −0.23 V. Is methane produced under these conditions in detectable amounts?

Solution

1. $pe = pe^0 (W) + 1/8 (\log pCO_2/pCH_4) = -4.13 + 0.125 \log (100/250) = -4.18$
2. $EH = 0.05895*(-4.18) = -0.246$
3. $-4.18 = 13.0 + \log Fe^{3+}/Fe^{2+}$; $[Fe^{3+}]/[Fe^{2+}] = 10^{-17.2}$
4. $0.125 \log (10^{-7.6}/10^{-7.0})^8 = -0.6$
5. The redox potential implies that $1/8 (\log pCO_2/pCH_4) \gg 0$

EXAMPLE 7.12

In the lagoon of Venice, sulfate may be transformed to free sulfur in hot weather with no or little wind. pe for this redox reaction is 6.03.

1. What is pe under these conditions? pH is measured to be 7.5. A 10 mM concentration of sulfate is assumed.
2. What is the ratio $[Fe^{3+}]/[Fe^{2+}]$ under these conditions? pe^0 for Fe^{3+}/Fe^{2+} is 13.0.

Solution

1. $pe = 6.03 + \log [SO_4^{2-}]^{1/6} [H^+]^{4/3} = 6.03 + (-0.33-10) = -4.3$
2. $-4.3 = 13.0 + \log [Fe(III)]/[Fe(II)]$ $[Fe(III)]/[Fe(II)] = 10^{-17.3}$

Figure 7.21, named a redox staircase, can be used in the same manner to decide on which oxidants are able to oxidize which reductants. pe^0 values for various redox pair at four different pH are shown. The oxidants placed higher in the figure can oxidize the reductants placed lower in the figure. At the same time, the figure indicates the pe value corresponding to an equilibrium of a given redox pair. If we know, for instance, that oxygen is present at a normal concentration around 8–10 mg/L in an aquatic ecosystem (dependent on the temperature), it will inevitably imply that pe must be around 13.75 and that the other redox pair must adjust their ratio according to this pe value. Note that pe^0 for Fe^{3+}/Fe^{2+} is independent of pH because no protons participate in the redox reaction.

Finally, the redox staircase indicates which oxidant will be used for an oxidation under given circumstances. If pe is around 13–15 and pH = 7.0, oxygen will be applied as oxidant before iron(III), nitrate, etc. When oxygen has been used, Fe^{3+} will be used before nitrate and sulfate, etc. This is parallel to the titration of a mixture of acids with a base. The strongest acid will be neutralized first, then the second strongest acid, etc. It is interesting that the sequence of oxidants (oxygen is used as oxidant before iron(III), which is used before nitrate, which is used before sulfate, which is used before carbon dioxide) also is the sequence that gives the highest energy efficiency (most kJ for formation of ATP and thereby most exergy; see also Jørgensen, 2012). This is shown in Table 7.7.

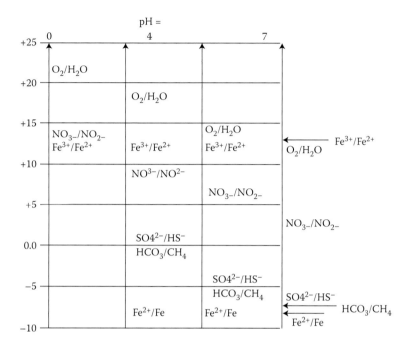

FIGURE 7.21

pe° values for the most common redox pair in aquatic systems for four different pH values.

TABLE 7.7

Yields of kJ and ATPs per Mole of Electrons Corresponding to 0.25 Moles of CH_2O Oxidized

Reaction	kJ/mol e⁻	ATPs/mol e⁻
$CH_2O + O_2 = CO_2 + H_2O$	125	2.98
$CH_2O + 0.8NO_3^- + 0.8H^+ = CO_2 + 0.4N_2 + 1.4H_2O$	119	2.83
$CH_2O + 2MnO_2 + H^+ = CO_2 + 2Mn^{2+} + 3H_2O$	85	2.02
$CH_2O + 4FeOOH + 8H^+ = CO_2 + 7H_2O + Fe^{2+}$	27	0.64
$CH_2O + 0.5SO_4^{2-} + 0.5H^+ = CO_2 + 0.5HS^- + H_2O$	26	0.62
$CH_2O + 0.5CO_2 = CO_2 + 0.5CH_4$	23	0.55

Note: The released energy is available to build ATP for various oxidation processes of organic matter at pH = 7.0 and 25°C.

7.24 Construction of pe–pH Diagrams

The equations for pe obtained from Equations 7.80 and 7.81

$$pe = -\,pH - 0.5 \log pH_2 \qquad (7.92)$$

$$pe = 20.78 - pH + 0.25 \log pO_2 \qquad (7.93)$$

can be plotted in a pH–pe diagram (see Figure 7.22). The diagram gives the following important information: above the upper line, water is an effective reductant (forming oxygen) and below the lower line, water is an effective oxidant, producing hydrogen. Aquatic systems in equilibrium with the oxygen in the atmosphere will have a pe = f(pH) as the upper line. Under these normal aerobic conditions, oxygen is an oxidant and hydrogen is a reductant.

pe–pH diagrams illustrate which components will prevail under given conditions, that is, at a given pe and pH. It is therefore advantageous to construct pe–pH diagrams for the most commonly found elements in aquatic systems to determine quickly which form of the considered element is stable under the prevailing conditions. At the same time, the pe–pH diagram gives

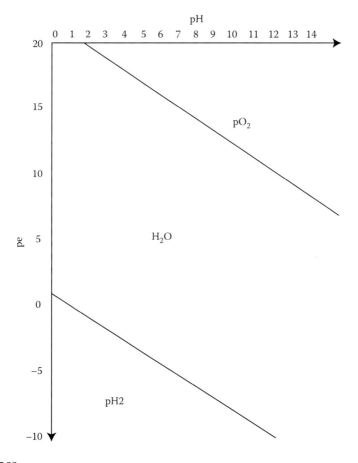

FIGURE 7.22
The equilibria between water and oxygen (upper line) and between water and hydrogen (lower line).

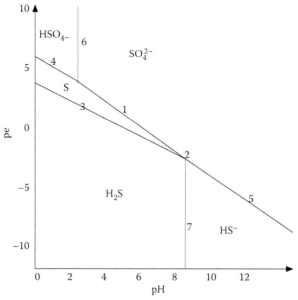

1: pe = 6.03 + 1/6 log [SO4²⁻] −8/6 pH   2: pe = 8.01 + 1/8 log [SO4²⁻] / [H₂S] −10/8 pH
3: pe = 2.4 − pH − 0.5 log [H₂S]   4: pe = 5.7 + 1/6 log [HSO4⁻] − 7/8 pH
5: pe = 4.25 + 1/8 log [SO4²⁻] / [HS⁻] −9/8 pH
6: log [SO4²⁻] / [HSO4⁻] −pH = 2.0   7: log [HS⁻] / [H₂S] /−pH = 7.0

FIGURE 7.23
pe–pH diagram for the sulfate–sulfur–hydrogen sulfide system.

a good overview of the possible processes where both redox processes and acid–base reactions can take place. Figure 7.23 gives the information for the two elements, oxygen and hydrogen. Example 7.13 gives the pe–pH diagram for the sulfur system.

EXAMPLE 7.13

Construct a diagram considering that sulfur can be found in natural aquatic ecosystem as sulfate, sulfur (s), and hydrogen sulfide (g) in equilibrium with aqueous solution of hydrogen sulfide. The following redox reactions can be found in the literature:

$$SO_4^{2-} + 8H^+ + 6e^- = S(s) + 4H_2O \quad pe^0 = 6.03$$

$$SO_4^{2-} + 10H^+ + 8e^- = H_2S(aq) + 4H_2O \quad pe^0 = 5.12$$

$$S(s) + 2H^+ + 2e^- = H_2S(aq) \quad pe^0 = 2.40$$

$$HSO_4^- + 7H^+ + 6e^- = S(s) + 4H_2O \quad pe^0 = 5.70$$

$$SO_4^{2-} + 9H^+ + 8e^- = HS^- + 4H_2O \quad pe^0 = 4.25$$

Hydrogen sulfate has $pK = 2.0$ and hydrogen sulfide has $pK = 7.0$. The total concentration of the soluble sulfur species is 0.01 M.

Solution

Based on the shown reactions, it is possible to obtain the following equations for the pe–pH diagram:

$$pe = 6.03 + 1/6 \cdot \log [SO_4^{2-}] - 8/6 \cdot pH$$

$$pe = 5.12 + 1/8 \cdot \log ([SO_4^{2-}]/[H_2S(aq)]) - 10/6 \cdot pH$$

$$pe = 2.40 - 1/2 \cdot \log [H_2S] - pH$$

$$pe = 5.70 + 1/6 \cdot \log [HSO_4^{2-}] - 876 \cdot pH$$

$$pe = 4.25 + 1/8 \cdot \log ([SO_4^{2-}]/[H_2S(aq)]) - 9/8 \cdot pH$$

$$\log ([SO_4^{2-}]/[HSO_4^{2-}]/[H_2S(aq)]) - pH = -2.0$$

$$\log ([HS^-]/[H_2S(aq)]) - pH = -7.0$$

These equations can easily be plotted in a pe–pH diagram (see Figure 7.23).

If only one soluble species is included in the equation, the concentration 0.01 M is applied. If two soluble species are included, for instance, equation number 2, they are both assumed as 0.00 M. The resulting diagram is shown in Figure 9.5.

Figure 7.24 shows another example, namely, the pe–pH diagram for iron. It is constructed by the same method as Figures 7.22 and 7.23.

7.25 Redox Potential and Complex Formation

Figure 7.25 shows how it is possible to calculate the pe° value for two different oxidation states of metal complexes. It is possible to go from Me^{2+} to MeL^{2+} by the processes (1) + (3) or by the processes (2) + (4) and these two pathways must necessarily give the same result with respect to the overall equilibrium between Me^{2+} and MeL^{2+}. The equilibrium constants for the two pathways are therefore equal. The equilibrium constant for two successive processes are the product of the equilibrium constants for the two processes:

$$n\,pe^0 + \log K = \log K' + n\,pe'^0 \tag{7.94}$$

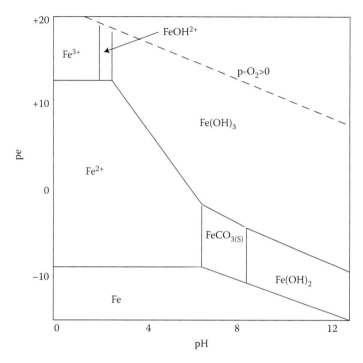

FIGURE 7.24
pe–pH diagram for iron.

As seen from this equation, if the highest oxidation state has the strongest complex, pe° will be reduced by addition of the corresponding ligand, and vice versa; if the lowest oxidation state has the strongest complex, the pe° will be increased by addition of the ligand.

$$Me^{2+} \xrightarrow[\ (2)\ ]{+L^-\ K} MeL^+$$

pe° ↓↑ (1)   (4) ↓↑ pe'o?

$$Me^{3+} \xrightarrow[\ (3)\ ]{+L^-\ K'} MeL^{2+}$$

$n\ pe^0 + \log K = \log K' + n\ pe^0$
When pe^0, log K and log K'
are known, pe^0 can be determined

FIGURE 7.25
(**See color insert.**) It is possible to go from Me^{3+} to MeL^{2+} by the processes (1) + (3) or by the processes (2) + (4) and the two pathways must necessarily give the same result with respect to the equilibrium between Me^{3+} and MeL^{2+}. It implies that $n\ pe^0 + \log K = \log K' + n\ pe^0$.

EXAMPLE 7.14

The complexity constants for formation of Fe^{3+}-hydroxo complexes by hydrolysis at room temperature are for mono-, di-, tri-, and tetra-hydroxo complexes, the so-called *β values, $10^{-2.2}$, $10^{-5.7}$, $10^{-15.6}$, and $10^{-21.6}$, respectively.

The complexity constants for the formation of fluoride complexes for Fe^{3+} at room temperature are

$$Fe^{3+} + F^- = FeF^{2+} \quad \log \beta_{1F} = 5.2$$

$$Fe^{3+} + 2F^- = FeF_2^+ \quad \log \beta_{2F} = 9.2$$

$$Fe^{3+} + 3F^- = FeF_3^0 \quad \log \beta_{3F} = 11.9$$

Iron(II) does not form complexes with fluoride and the possible hydroxo complexes are much weaker than the corresponding complexes for Fe^{3+}, and they can therefore be neglected. The pe° for the process

$$Fe^{3+} + e^- = Fe^{2+}$$

is 13.00 at room temperature.

A. Find the pe value at equal concentrations of iron(II) and iron(III), pH = 5.0, and a fluoride concentration of 0.01 M.
B. Which iron(III) complexes are strongest by these conditions?

Solution

A.

$$\alpha_0 = [Fe^{3+}] / Fe(III)$$

$$= 1/(1 + *\beta_1/[H^+] + *\beta_2/[H^+]^2 + *\beta_3/[H^+]^3 + *\beta_4/[H^+]^4$$

$$+ \beta_{1F}[F^-] + \beta_{2F}[F^-]^2 + \beta_{3F}[F^-]^3)$$

$$= 1/(1 + 10^{2.8} + 10^{4.3} + 10^{-0.6} + 10^{-1.6} + 10^{3.2} + 10^{5.2} + 10^{5.9})$$

$$= 1/9.7 \cdot 10^5 = 1.03 \cdot 10^{-6}$$

$$pe = 13.00 + \log [Fe^{3+}]/[Fe^{2+}] = 13.00 + \log 10^{-6}/1 = 7.00$$

B. From the expression for $\alpha_0 = [Fe^{3+}]/Fe(III)$, it can be seen that the dihydroxo, difluoride, and trifluoride complexes are contributing most to the formation of complexes.

7.26 Summary and Conclusions

It is important in integrated environmental management to assess the environmental concentrations of all possible compounds that can be a result of the chemical transformations in the environment. It will in most cases require an extensive use of chemical calculations. A number of useful chemical calculations have been presented in this chapter. If we know the discharge of various chemical compounds and the compositions of the ecosystems receiving the discharge, we can perform the needed chemical calculations.

References

Alloway, B.J. and D.C. Ayres. 1993. *Chemical Principles of Environmental Pollution.* Blackie Academic and Professional, London, 394pp.

Geyer, H. et al. Estimation of the biological concentration factor. *Chemosphere* 11: 1121–1130.

Jørgensen, S.E. 2000. *Principles of Pollution Abatement.* Elsevier, Amsterdam, 520pp.

Jørgensen, S.E. 2011. *Fundamentals of Systems Ecology.* CRC Press, Boca Raton, 320pp.

Jørgensen, S.E. 2012. *Fundamentals of Systems Ecology.* CRC Press, Boca Raton, 320pp.

Jørgensen, S.E., S. Nors Nielsen, and L.A. Jørgensen. 1991. *Handbook of Ecological Parameters and Ecotoxicology.* Elsevier, Amsterdam. Published as CD under the name ECOTOX, with L.A. Jørgensen as first editor in year 2000.

Jørgensen, S.E., B. Halling-Sørensen, and H. Mahler. 1997. *Handbook of Estimation Methods in Ecotoxicology and Environmental Chemistry.* Lewis Publishers, Boca Raton, 230pp.

Jørgensen, S.E. and G. Bendoricchio. 2001. *Fundamentals of Ecological Modelling* (3rd edition). Elsevier, Amsterdam, 628pp.

Leeuwen, C.J. van, and J.L.M. Hermens. 1995. *Risk Assessment of Chemicals: An Introduction.* Kluwer Academic Publishers, Dordrecht, 288pp.

Sillen, L.G. 1961. Oceanography Publication Number 67. AAAS, Washington D.C., pp. 549–581.

Weber, W.J. and F.A. DiGiano. 1996. *Process Dynamics in Environmental Systems.* John Wiley and Sons Inc., New York, 942pp.

8

Environmental Risk Assessment

8.1 Environmental Risk Analysis

This chapter provides a brief introduction to the concepts of environmental risk assessment (ERA). It aims to present the readers with information as to how to combine the knowledge about concentration of a polluting component with the properties of the component to assess the environmental risk. ERA may be considered a diagnostic tool. The concentration in the environment of a polluting component is usually known by analyses, or it can be estimated or be found more or less indirectly by environmental modeling (see, for instance, Jørgensen and Fath, 2011 and Chapter 9). The properties of the polluting components that are usually more difficult to assess than the EECs have been covered in Chapters 3 and 6, and methods to estimate the properties have been presented in Chapter 4. An integration of all sources and their effects is needed before it is possible to apply the diagnostic tools that are discussed in Chapters 9 and 10. This section introduces the idea and basic concept of ERA, while the last section presents ERA step by step.

An ERA is an analysis leading to determination of a risk of adverse effects on the environment or on human health. It is therefore a diagnostic tool. Most often the adverse effect is considered to be caused by a toxic chemical component, but its principle and the procedure can be used to any impact of a specific component on an ecosystem. Environmental exposure concentrations are determined or predicted and compared with concentrations in different environmental compartment of no-effect concentrations. ERA reveals whether measures are needed to limit the potential environmental consequences of a substance, and it can furthermore be pointed out which further testing and knowledge are needed to perform the ERA. ERA is based on the same basic concepts as the environmental impact assessment (EIA), which includes the use of the diagnosis tools presented in the next chapters. Some EIAs have even included the selection of proper solutions tools (Chapters 11 through 13) and sometimes even the final integration of all the steps (presented in Chapter 14). Both ERA and EIA use models, indicators, and ecosystem services as diagnostic tools, but EIA is usually integrated with the diagnosis to a higher extent, while ERA in some cases may be performed independently of the diagnosis. Both ERA and EIA are, however, important tools in integrated environmental management.

Treatment of industrial air pollutants and wastewater, more or less toxic solid waste, and smoke is often very expensive. Consequently, industries attempt to change their products and production methods in a more environmentally friendly direction to reduce the treatment costs. Industries need, therefore, to know how much the different chemicals, components, and processes are polluting our environment. Or expressed differently: What is the environmental risk of using a specific material or chemical compared with other alternatives? If industries can reduce their pollution just by switching to another chemical or process, they will consider doing so to reduce their environmental costs or improve their green image. An assessment of the environmental risk associated with the use of a specific chemical and a specific process gives industries the possibility of making the right selection of materials, chemicals, and processes to the benefit of the economy of the enterprise and the quality of the environment.

Similarly, society needs to know the environmental risks of all chemicals used in society to phase out the most environmentally threatening chemicals and set standards for the use of all other chemicals. The standards should ensure that there is no serious risk in using the chemicals, provided that the applied standards are followed carefully. Modern abatement of pollution therefore includes ERA, which may be defined as the process of assigning magnitudes and probabilities to the adverse effects of human activities. The process involves identification of hazards such as the release of toxic chemicals to the environment by quantifying the relationship between an activity associated with an emission to the environment and its effects. The entire ecological hierarchy is considered in this context including the effects on the cellular (biochemical) level, on the organism level, on the population level, on the ecosystem level, and for the entire ecosphere.

The application of ERA is rooted in the recognition that

1. The elimination cost of all environmental effects is impossibly high
2. Practical environmental management decisions must always be made on the basis of incomplete information

As already mentioned, we use about 100,000 chemicals in such amounts that they might threaten the environment, but we know about only 1% of what we need to know to make a proper and complete ERA of all these chemicals. Chapter 4 presents a number of estimation methods that can be applied if we cannot find information about properties of chemical compounds in the literature. A list of the relevant properties are touched upon in Chapters 3, 4, and 6 and it is discussed in this chapter as well; what these properties mean for the environmental impact—or expressed differently the properties are translated to a certain extent into possible environmental effects.

As mentioned above, ERA is in the same family as EIA, which attempts to assess the impact of a specific human activity. EIA is predictive, comparative, and concerned with all possible effects on the environment, including secondary and tertiary (indirect) effects, while ERA attempts to assess the probability of a given (defined) adverse effect as a result of human activity and/or the use of a specific chemical component.

Both ERA and EIA use models to find the expected EEC, which is translated into impacts for EIA and to risks of specific effects for ERA. An overview of ecotoxicological models is given in Jørgensen and Fath (2011) and will furthermore be presented in Chapter 9.

Legislation and regulation of domestic and industrial chemicals with respect to the protection of the environment have been implemented in Europe and North America for decades. Both regions distinguish between existing chemicals and introduction of new substances. For existing chemicals, the European Union requires a risk assessment to humans and environment according to a priority setting. An informal priority setting (IPS) is used for selecting chemicals among the 100,000 listed in the European Inventory of Existing Commercial Chemical Substances. The purpose of IPS is to select chemicals for detailed risk assessment from among the EEC high production volume compounds, that is, >1000 t/year (about 2500 chemicals). Data necessary for the IPS and an initial hazard assessment are called Hedset and cover issues as environmental exposure, environmental effects, exposure to humans, and human health effects. The latest European development, the Reach Reform, is presented in Section 13.7.

At the United Nations Conference on Environment and Development (UNCED) meeting on environment and sustainable development in Rio de Janeiro in 1992, it was decided to create an Intergovernmental Forum on Chemical Safety (IGFCS, Chapter 19 of Agenda 21). The primary task of IGFCS is to stimulate and coordinate global harmonization in the field of chemical safety, covering the following principal themes: assessment of chemical risks, global harmonization of classification and labeling, information exchange, risk reduction programs, and capacity building in chemicals management.

Uncertainty plays an important role in risk assessment (Suter, 1993). Risk is the probability that a specified harmful effect will occur or in the case of a graded effect, the relationship between the magnitude of the effect and its probability of occurrence.

Risk assessment has often emphasized risks to human health but should under no circumstances ignore ecological effects. Some chemicals have no or only little risk to human health but may cause severe effects on ecosystems such as aquatic organisms. Examples are chlorine, ammonia, and certain pesticides. An up-to-date risk assessment comprises considerations of the entire ecological hierarchy, which is the ecologist's worldview in terms of levels of organization. Organisms interact directly with the environment and are exposed to toxic chemicals. The species–sensitivity distribution is therefore more ecologically credible (Calow, 1998). A reproducing population is the smallest meaningful level in an ecological sense. However, populations do not exist in vacuum but require a community of other organisms of which the population is a part. The community occupies a physical environment with which it forms an ecosystem.

Moreover, both the various adverse effects and the ecological hierarchy have different scales in time and space, which must be included in a proper ERA (see Figure 8.1), where log spatial (indicated as length) scales of various hazards versus log of the time scale are plotted to recover after the hazard and for ecological processes at different levels of the ecological hierarchy. For example, oil spills occur at a spatial scale similar to those of population dynamics, but they are briefer than population processes. Therefore,

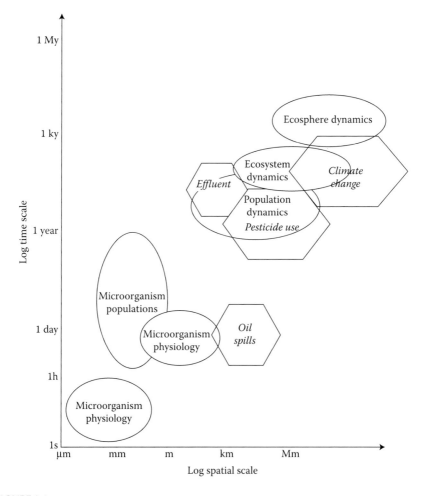

FIGURE 8.1
The spatial (ads length) and time scale for various hazards (hexagons, italic) and for the various levels of the ecological hierarchy (circles, nonitalic).

a risk assessment of an oil spill requires considering reproduction and recolonization on a longer time scale to determine the magnitude of the population response and its significance to natural population variance.

EXAMPLE 8.1

The time to recover after an oil spill with

A. A damage area of 100×100 km in the tropic climate requires 5 years
B. A damage area of 200×200 km requires in the tropic climate requires 10 years
C. A damage area of 100×100 km in the arctic climate 40 years

Explain this difference.

Solution

The recovery time is a function of the circumvent relative to the area and the temperature. The circumvent in case B relative to the area is half the value in case A. It means that the recovery processes (ecological processes) in case B are twice the recovery processes in case A, but as four times as much area has to recover it will take twice as long time in case B as in case A.

The temperature in case C is presumed about 30°C lower than in case A, and the rate of ecological processes double for every 10°C increase in temperature. It means that the ecological processes will have half the rate at 10°C decrease in temperature; one quarter the rate at 20°C decrease in temperature and eight times lower rate at a temperature of 30°C lower. Therefore, case C will require 5 years' times $8 = 40$ years!

Uncertainties in risk assessment are taken most commonly into account by application of safety factors. Uncertainties have three basic causes:

1. The inherent randomness of the world (stochasticity)
2. Errors in execution of assessment
3. Imperfect or incomplete knowledge

The inherent randomness refers to uncertainty that can be described and estimated but cannot be reduced because it is characteristic of the system. Meteorological factors such as rainfall, temperature, and wind are effectively stochastic at levels of interest for risk assessment. Many biological processes such as colonization, reproduction, and mortality also need to be described stochastically.

Human errors are inevitably attributes of all human activities. This type of uncertainty includes incorrect measurements, data recording errors, computational errors, and so on. Uncertainty is addressed using an assessment (safety) factor from 10 to 1000. The choice of assessment factor depends on the quantity and quality of toxicity data (see Table 8.1). The assessment or safety factor is used in Step 3 of the ERA procedure presented below. Other relationships than the uncertainties originating from randomness, errors, and lack of knowledge may be considered when the assessment factors are selected, for instance, cost–benefit. This implies that the assessment factors for drugs and pesticides may be given a lower value due to their possible benefits.

Lack of knowledge results in undefined uncertainty that cannot be described or quantified. It is a result of practical constraints on our ability to accurately describe, count, measure, or quantify everything that pertains to a risk estimate. Clear examples are the inability to test all toxicological responses of all species exposed to a pollutant and the simplifications needed in the model used to predict the expected EEC.

8.2 Development of ERA

The most important feature distinguishing risk assessment from impact assessment is the emphasis in risk assessment on characterizing and quantifying uncertainty. Therefore, it is of particular interest in risk assessment to analyze and estimate the analyzable uncertainties. They are natural stochasticity, parameter errors, and model errors. Statistical methods may provide direct estimates of uncertainties. They are widely used in model development.

The use of statistics to quantify uncertainty is complicated in practice by the needs to consider errors in both the dependent and independent variables and to combine errors when multiple extrapolations should be made. Monte Carlo analysis is often used to overcome these difficulties (see, for instance, Bartell et al., 1992). Model errors include inappropriate selection or aggregation of variables, incorrect functional forms, and incorrect boundaries. The uncertainty associated with model errors is usually assessed by field measurements utilized for calibration and validation of the model (see Chapter 9). The result of the validation is a determination of the model uncertainty. The modeling uncertainty for ecotoxicological models is in principle not different from other models, as stated in the modeling literature (see Jørgensen and Fath, 2011).

Chemical risk assessment may be divided into nine steps, which are shown in Figure 8.2. The nine steps correspond to questions that the risk assessment attempts to answer to quantify the risk associated with the use of a chemical. The nine steps are presented in detail below with reference to Figure 8.2.

Step 1: Which hazards are associated with the application of the chemical? This involves gathering data on the types of hazards—possible environmental damage and human health effects. The health effects include congenital, neurological, mutagenic, endocrine disruption (so-called estrogen effect), and carcinogenic effects. It may also include characterization of the behavior of the chemical within the body (interactions with organs, cells, or genetic material). What is the possible environmental damage including lethal effects and sublethal effects on growth and reproduction of various populations.

As an attempt to quantify the potential danger posed by chemicals, a variety of toxicity tests has been devised. Some of the recommended tests involve experiments with subsets of natural systems, for instance, microcosms or with entire ecosystems. The majority of testing new chemicals for possible effects has, however, been confined to studies in the laboratory on a limited number of test species. Results from these laboratory assays provide useful information for quantification of the relative toxicity of different chemicals. They are used to forecast effects in natural systems, although their justification has been seriously questioned (Cairns et al., 1987).

Step 2: What is the relation between dose and response of the type defined in Step 1? It implies knowledge of NEC (noneffect concentration), LD_x (the dose which is lethal to x% of the organisms considered), LC_y (the concentration which is lethal to y% of the organisms considered), and EC_z values (the concentration giving the indicated effect to z% of the considered organisms), where x, y, and z express a probability of harm. The answer can be found by laboratory examination or we may use estimation methods. Based on these answers, a most probable level of no effect, NEL, is assessed. Data required for Steps 1 and 2 can be obtained directly from scientific libraries, but are increasingly found via online data searches in bibliographic and factual databases. Data gaps should be filled with estimated data. It is very difficult to obtain complete knowledge about the effect of a chemical on all levels from cells to ecosystem. Some effects are associated with very small concentrations, such as the estrogen effect. It is therefore far from sufficient to know NEC, LD_x, LC_y, and EC_z values.

Step 3: Which uncertainty (safety) factors reflect the amount of uncertainty that must be taken into account when experimental laboratory data or empirical estimation methods are extrapolated to real situations? Usually, safety factors of 10–1000 are used. The choice is discussed earlier and will be in accordance with Table 8.1. If good knowledge about the chemical is available, then a safety factor of 10 may be applied. If, on the other hand, it is estimated that the available information has a very high uncertainty, then a safety factor of 10,000 may be recommended. Most frequently, safety factors of 50–100 are applied. NEL times the safety factor is named the predicted noneffect level, PNEL. The complexity of ERA is often simplified by deriving the predicted no effect concentration, PNEC, for different environmental components (water, soil, air, biotas, and sediment).

Step 4: What are the sources and quantities of emissions? The answer requires thorough knowledge of the production and use of the chemical compounds considered, including an assessment of how much of the chemical is

TABLE 8.1

Selection of Assessment Factors to Derive PNEC[a]

Data Quantity and Quality	Assessment Factor
At least one short-term LC_{50} from each of the three trophic levels of the base set (fish, zooplankton, and algae)	1000
One long-term NOEC (nonobserved effect concentration, for either fish or *Daphnia*)	100
Two long-term NOECs from species representing two trophic levels	50
Long-term NOECs from at least three species (normally fish, *Daphnia*, and algae) representing three trophic levels	10
Field data or model ecosystems	Case by case

[a] See also Step 3 of the procedure.

wasted in the environment by production and use? The chemical may also be a waste product, which makes it very difficult to determine the amounts involved. For instance, the very toxic dioxins are waste products from incineration of organic waste under certain conditions.

Step 5: What is (are) the actual exposure concentration(s)? The answer to this question is called the predicted environmental concentration (PEC). Exposure can be assessed by measuring EECs. It may also be predicted by a model when the emissions are known. The use of models is necessary in most cases either because we are considering a new chemical or because the assessment of EECs requires a very large number of measurements to determine the variations in concentrations in time and space. Furthermore, it provides an additional certainty to compare model results with measurements, which imply that it is always recommended to develop a model and make at least a few measurements of concentrations in the ecosystem components when and where it is expected that the highest concentration will occur. Most models will demand an input of parameters, describing the properties of the chemicals and the organisms, which also will require extensive application of handbooks and a wide range of estimation methods. The development of an environmental, ecotoxicological model requires extensive knowledge of the physical–chemical–biological properties of the chemical compound(s) considered. The selection of a proper model is discussed in Chapter 9 and in Jørgensen and Fath (2011).

Step 6: What is the ratio PEC/PNEC? Often called the risk quotient, the ratio should not be considered an absolute assessment of risk but rather a relative ranking of risks. The ratio is usually found for a wide range of ecosystems such as aquatic ecosystems, terrestrial ecosystems, and groundwater.

Steps 1–6, shown in Figure 8.3, agree with Figure 8.2 and the information given previously.

Step 7: How will you classify the risk? Risk valuation is made to decide on risk reductions (Step 9). Two risk levels are defined:

1. The upper limit, that is, the maximum permissible level (MPL).
2. The lower limit, that is, the negligible level (NL). It may also be defined as a percentage of MPL, for instance, 1% or 10% of MPL.

The two risk limits create three zones: a black, unacceptable, high-risk zone > MPL, a gray, medium-risk level, and a white, low-risk level < NL. The risk of chemicals in the gray and black zones must be reduced. If the risk of the chemicals in the black zone cannot be reduced sufficiently, then it should be considered to phase out the use of these chemicals.

Step 8: What is the relation between risk and benefit? This analysis involves examination of socioeconomic, political, and technical factors, which are beyond the scope of this volume. The cost–benefit analysis is difficult because the costs and benefits are often of a different order and dimension.

Step 9: How can the risk be reduced to an acceptable level? The answer to this question requires a deep technical, economic, and legislative investigation. Assessment of alternatives is often an important aspect in risk reduction.

Steps 1–3 and 5 require knowledge of the properties of the focal chemical compounds, which again implies an extensive literature search and/or selection of the best feasible estimation procedure. In addition to "Beilstein" it can be recommended to have at hand the following very useful handbooks of environmental properties of chemicals and methods for estimation of these properties in case literature values are not available:

Jørgensen, S.E., S.N. Nielsen, and L.A. Jørgensen. 1991. *Handbook of Ecological Parameters and Ecotoxicology*. Elsevier, 1991. Year 2000 published as a CD called Ecotox. It contains three times the amount of parameter in the 1991 book edition. See also Chapter 2 for further details about Ecotox.

Howard, P.H. et al. 1991. *Handbook of Environmental Degradation Rates*. Lewis Publishers.

Verschueren, K. Several editions have been published the latest in 2007. *Handbook of Environmental Data on Organic Chemicals*. Van Nostrand Reinhold.

Mackay, D., W.Y. Shiu, and K.C. Ma. *Illustrated Handbook of Physical–Chemical Properties and Environmental Fate for Organic Chemicals*. Lewis Publishers.

Volume I. *Monoaromatic Hydrocarbons. Chlorobenzenes and PCBs*. 1991.

Volume II. *Polynuclear Aromatic Hydrocarbons, Polychlorinated Dioxins, and Dibenzofurans*. 1992.

Volume III. *Volatile Organic Chemicals*. 1992.

If the literature cannot give you the answer about the needed properties of a chemical for development of ERA, it will be important to use QSAR;

see Chapter 4 and Jørgensen, S.E., Mahler, H., and Halling-Sørensen, B., 1997. *Handbook of Estimation Methods in Environmental Chemistry and Ecotoxicology.* Lewis Publishers.

Steps 1–3 are sometimes denoted as effect assessment or effect analysis and Steps 4–5 exposure assessment or effect analysis. Steps 1–6 may be called risk identification, while ERA encompasses all the nine steps presented in Figure 8.2. Particularly, Step 9 is very demanding, as several possible steps

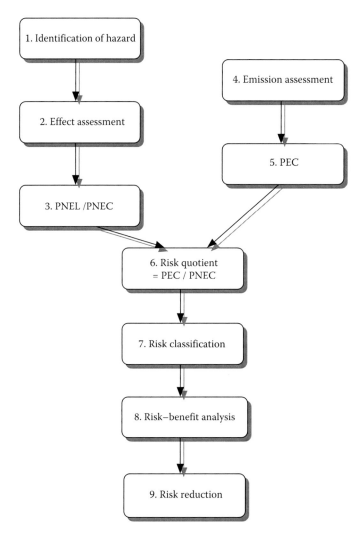

FIGURE 8.2
The presented procedure in nine steps to assess the risk of chemical compounds. Steps 1–3 require extensive use of ecotoxicological handbooks and ecotoxicological estimation methods to assess the toxicological properties of the chemical compounds considered, while Step 5 requires the selection of a proper ecotoxicological model.

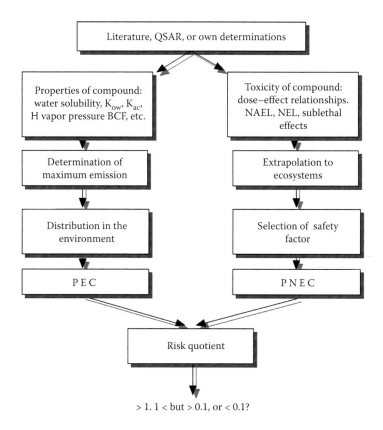

FIGURE 8.3
Steps 1–6 shown in more detail for practical applications. The result of these steps naturally leads to the assessment of the risk quotient.

in reduction of the risk should be considered, including treatment methods, cleaner technology, and substitutes to replace the examined chemical. In North America, Japan, and EU, medicinal products are considered similar to other chemical products, as there is no difference in principle between medicinal products and other chemical products. It is also possible to perform an ERA where the human population is in focus. The 10 steps corresponding to Figure 8.3 are shown in Figure 8.4, which is not very significantly different from Figure 8.3. The principles for the two types of ERA are the same. Figure 8.4 uses the nonadverse effect level (NAEL) and nonobserved adverse effect level (NOAEL) to replace the predicted noneffect concentration, and the predicted EEC is replaced by the tolerable daily intake (TDI).

This type of ERA has a particular interest for veterinary medicine, which may contaminate food products for human consumption. For instance, the use of antibiotics in pig feed has attracted considerable attention, as they may be found as residue in pig meat or may contaminate the environment through the application of manure as a natural fertilizer.

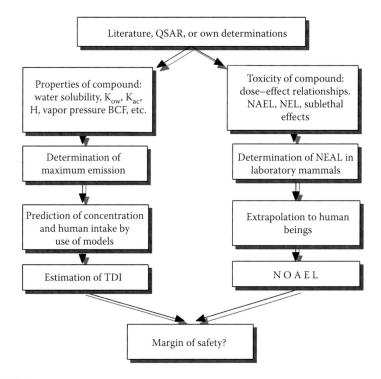

FIGURE 8.4
ERA for human exposure. It leads to a margin of safety, which corresponds to the risk quotient in Figures 8.2 and 8.3.

Selection of a proper ecotoxicological model is a first initial step in the development of an environmental exposure model, as required in Step 5. It will be discussed in more detail in the next chapter.

References

Bartell, S.M., R.H. Gardner, and R.V. O'Neill. 1992. *Ecological Risk Estimation.* Lewis Publishers, Boca Raton.

Cairns, Jr., J., K.L. Dickson, and A.W. Maki. 1987. *Estimating Hazards of Chemicals to Aquatic Life.* STP. 675, American Society for Testing and Materials, Philadelphia.

Calow, P. 1998. Ecological risk assessment: Risk for what? How do we decide? *Ecotoxicology and Environmental Safety*, 40: 15–18.

Jørgensen, S.E. and B. Fath. 2011. *Fundamentals of Ecological Modeling.* 4th Edition. Elsevier, Amsterdam, 400pp.

Jørgensen, S.E., B. Halling-Sørensen, and H. Mahler. 1997. *Handbook of Estimation Methods in Ecotoxicology and Environmental Chemistry.* Lewis Publishers, Boca Raton, 230pp.

Suter, G.W. 1993. *Ecological Risk Assessment.* Lewis Publishers, Chelsea.

9

Application of Ecological Models in Environmental Chemistry and Ecotoxicology

9.1 Physical and Mathematical Models

Mankind has always made more or less descriptive models of the environment, which often have been used to solve problems or to obtain knowledge about how nature is functioning. When we use mathematics as a useful tool in our description of nature, we talk about mathematical models in contrast to physical models, which try to imitate the physical appearance of the modeled object, for instance, a model of a ship. Models attempt to provide a simplified picture of reality and will never be able to contain all the features of the real system, because then it would be the real system itself. It is, however, important that the model contains the characteristics of the system that are essential in the context of the management or scientific problem to be solved or comprehended.

The philosophy behind the use of a model might be best illustrated by an example. For many years, we have used physical models of ships to determine the profile that gives a ship the smallest resistance in water. Such a model will have the shape and the relative main dimensions of the real ship, but will not contain all the details such as, for example, the instrumentation, the layout of the cabins, etc. Such details are of course irrelevant to the objectives of that model. Other models of the ship serve other aims: blue prints of the electrical wiring, layout of the various cabins, drawings of pipes, etc.

Correspondingly, an ecological model must contain the features that are of interest for the management or scientific problem that we wish to solve by the model. An ecosystem is a much more complex system than a ship. It implies that it is a far more complicated matter to capture the main features of importance for an ecological problem. However, intense research during the last few decades has made it possible today to set up many workable and applicable ecological models.

An ecological model focuses on the objects of interest for a considered well-defined problem. It would disturb the main objectives of a model to include too many irrelevant details. There are many different ecological models of the same ecosystem, as the model edition is selected according to the objectives of the model.

The model might be physical, such as the ship model used for the resistance measurements, which may be called microcosmos or it might be a mathematical model, which describes the main characteristics of the ecosystem and the related problems in mathematical terms.

Physical models will only be touched upon on very briefly in this chapter, which will focus almost entirely on mathematical models, particularly for application in environmental management. The field of ecological modeling has developed rapidly during the last three decades essentially due to three factors:

1. The development of computer technology, which has enabled us to handle very complex mathematical systems. In addition, the use of computer technology costs far less today than 30–40 years ago.

2. A general understanding of environmental problems, including that a complete elimination of pollution is not feasible (denoted zero discharge), but that a proper pollution control with limited economical resources available requires serious considerations of the influence of pollution impacts on ecosystems.

3. Our knowledge of environmental and ecological problems has increased significantly. We have particularly gained more knowledge on the quantitative relations in the ecosystems and between the ecological properties and the environmental factors.

Models may be considered a synthesis of what we know about the ecosystem with reference to the considered problem, in contrast to a statistical analysis, which only will reveal the relationships between the data. A model is able to include our entire knowledge about the system:

- Which components interact with which other components, for instance, zooplankton grazing on phytoplankton.
- Our knowledge about the processes often formulated as mathematical equations that have causality and been proved valid generally
- The importance of the various processes with reference to the problem

to mention a few examples of knowledge which may often be incorporated in an ecological model. This implies that a model can offer a deeper understanding of the entire system than a statistical analysis, which is often built on a correlation between a few sets of data, only. A carefully built ecological model is therefore a stronger tool in research than just data analysis and can result in a better management plan, on how to solve an environmental problem. This does not of course mean that statistical analytical results are not applied in the development of models. On the contrary, models are built on *all* available knowledge, including knowledge gained by statistical analyses of data, physical–chemical–ecological knowledge, the laws of nature, common sense, our general image on how nature is working, and so on. That is the great advantage of modeling.

9.2 Models as a Management Tool

The idea behind the use of ecological management models is demonstrated in Figure 1.1. Urbanization and technological development have had an increasing impact on the environment. Energy and pollutants are released into ecosystems, where they may cause more rapid growth of algae or bacteria, damage species, or alter the entire ecological structure. An ecosystem is extremely complex, and it is therefore an overwhelming task to predict the environmental effects that such emissions may have. It is here that the model comes into picture. With sound ecological knowledge, it is possible to extract the components and processes of the ecosystem that are particularly involved in a specific pollution problem to form the basis of the ecological model (see also the discussion in Chapter 2). As indicated in Figure 1.1, the model resulting can be used to select the environmental technology eliminating the emission most effectively.

Figure 1.1 represents the idea behind the introduction of ecological modeling as a management tool around the year 1970. The environmental management of today is more complex and applies therefore a wider spectrum of tools. Today, we have as alternative and supplement to environmental technology, cleaner technology, ecotechnology, environmental legislation, international agreements, and sustainable management plans. Ecotechnology is mainly applied to solve the problems of nonpoint or diffuse pollution, often originated from agriculture. The significance of nonpoint pollution was hardly acknowledged before 1980. Furthermore, global environmental problems play a much more important role today than 20 or 30 years ago, for instance, the reduction of the ozone layer and the climatic changes due to the greenhouse effect. The global problems can hardly be solved without international agreements and plans. Figure 1.2 attempts therefore to illustrate the more complex picture of environmental management today.

As mentioned above, the ecological systems are usually characterized by a much higher complexity than physical systems. Fortunately, we have today very powerful computers that can help us to develop complex models. Still, however, it is a crucial question in the development of ecological models: which are the important processes and components in the model context?

To answer this question, we need to know the problem and the ecosystem. Simplifications in calculations and in the development of ecological models require that we can distinguish the essential processes and components from the minor important ones, which should not be included in the model, because they only make the model unwieldy and difficult to overview.

Our problems are getting increasingly complex and highly comprehensive due to, among other factors, globalization. It is, therefore, becoming extremely difficult to survey all the factors that may influence the problems. Science is using an analysis of the problem and separate often in laboratory a problem and a few factors having influence on the problem; but many of the real-life problems are too complex to be solved by this scientific method. A synthesis of many factors and their interactions is needed and it implies that we have to develop a model to be able to survey the many possibilities. Some of the interactions and processes may be known from science; but for real-life problems, including environmental problems there are many interactions operating, simultaneously. Fortunately, models have been developed to deal with the high complexity and capture the characteristics and properties of systems. Today, we have a wide experience in the development of ecological models. Many thousands of models have been published during the last four decades, and the journal *Ecological Modelling* now publishes about 4000 pages per year. Ecological and environmental models are able to support our decisions in ecological and environmental management strongly.

Therefore, building a model helps us to perceive the interconnections and to connect previous experience with present observations and simulate what would happen in the future. Model development often requires a multidisciplinary approach because processes and interconnections from many disciplines may be an integrated part of the problem. The experience in the last several decades with the development of ecological models has shown us the clear advantages that models are able to offer, that is, ecology and environmental management today (Jørgensen and Fath, 2011):

- Models are synthesis of all what we know—observations, theoretical knowledge, knowledge about rates and sizes, knowledge about food items, etc.
- Models are tools to overview complex systems.
- Models make it possible to quantify by the use of mathematical formulations and computers.

It should be emphasized that all models may be considered a synthesis of all that we know about the problem and system. Models give a much more profound picture of the problem and the system than, for instance, a statistical treatment of observations because a model can include *all* our theoretical knowledge.

9.3 Modeling Components

Models have five types of components:

1. Forcing functions, or external variables, which are functions or variables of external impacts, inputs, and outputs that influence the state of the system. The problem to be solved can often be formulated as follows: if certain forcing functions vary, how will this variation influence the state of the ecosystem? The model is used to predict what will change in the system, when forcing functions are varied with time. The forcing functions under our control are often called control functions. The control functions in ecological models could, for instance, be inputs of toxic substances to the ecosystems and in eutrophication models the control functions are inputs of nutrients. Natural forcing functions in contrast to control functions are, for instance, the climatic variables, which influence the biotic and abiotic components and the process rates.

2. State variables describe, as the name indicates, the state of the system. The selection of state variables is crucial to the model structure, but often the choice is obvious. If, for instance, we want to model the bioaccumulation of a toxic substance in an ecosystem, the state variables should be the organisms in the most important food chains and concentrations of the toxic substance in the organisms. In eutrophication models, the state variables will be the concentrations of nutrients and phytoplankton. When the model is used in a management context, the values of state variables predicted by changing the forcing functions can be considered as the results of the model because the model contains relations between the forcing functions and the state variables.

3. Mathematical equations are used to represent the biological, chemical, and physical processes. The total set of equations describes the relationship between the forcing functions and state variables. The same type of processes may be found in many different environmental contexts, which imply that the same equations can be used in different models. This does not imply, however, that the same process is always formulated by use of the same equation. First, the considered process may be better described by another equation because of the influence of other factors. Second, the number of details needed or desired to be included in the model may be different from case to case due to a difference in complexity of the system and the problem. Some modelers refer to the description and mathematical formulation of processes as submodels. A comprehensive overview of submodels can be found in the ecological literature (see, for instance, *Handbook of Ecological Processes* by Palmeri et al., 2013).

4. Parameters are coefficients in the mathematical representation of processes. They may be considered constant for a specific ecosystem or part of an ecosystem, but probably in most cases dependent on the temperature. In causal models, the parameter will have a scientific definition, for instance, the excretion rate of a toxic substance from a fish (unit, for instance, mg/(g 24 h)). Most parameters are not indicated in the literature as constants but as ranges, but even that is of great value in the parameter estimation, as will be discussed further in the following text. In Jørgensen et al. (2000), a comprehensive collection of parameters in environmental sciences and ecology can be found. Our limited knowledge of parameters is one of the weakest points in ecological modeling, a point that will be touched upon often throughout the book. Chapter 4 reviews methods to estimate the properties of chemical compounds that are often needed as parameters in ecotoxicological models.

Furthermore, the applications of parameters as constants in our models are unrealistic due to the many feedbacks in real ecosystems. The flexibility of ecosystems is inconsistent with the application of constant parameters in the models. A new generation of models that attempt to use parameters varying according to some ecological principles seems a possible solution to the problem, but a further development in this direction is necessary before we can achieve an improved modeling procedure reflecting the processes in real ecosystems.

5. Universal constants, such as the gas constant and atomic weights, are also used in most models.

Mathematical models can be defined as formal expressions of the essential elements of a problem in mathematical terms. The first recognition of the problem is often verbal. This may be recognized as an essential preliminary step in the modeling procedure, which will be treated in more detail in the next section. The processes that are linking the state variables and the forcing functions are, however, easily expressed as mentioned above by the use of mathematical equations. In the most applied type of ecological models—dynamic biogeochemical models—differential equations are applied to express the changes of the state variables: accumulation = input − output according to the mass conservation principle. Ecotoxicological models are often biogeochemical models, where the conservation principles are used on the toxic compounds. These equations express simply that accumulation = inputs − outputs, where inputs and outputs are processes that are expressed by algebraic equations.

Many ecological researchers have discovered how powerful ecological models are to test scientific hypothesis about ecosystem reactions. Ecosystems are complex systems and it is therefore in most cases not easy to make direct experiments with entire ecosystems. It is, however, possible to make changes, for instance, in the forcing functions and observe the reaction of the ecosystems as changes of the state variables. In such cases, it will be necessary to develop a model of the focal ecosystem with the involved forcing functions and state variables to be able to assess whether the observations are in accordance with a proposed theory or hypothesis. These possibilities have opened up for a wider use of ecological models in ecology, particularly system ecology, of course.

9.4 Recommended Modeling Procedure

Figure 9.1 presents a procedure that is often applied in model development. The procedure can be considered the experience gained by the development of several different ecological models during the 1970s. The open question at that time was how can we ensure that the model is represented as correctly as it is possible the integrated knowledge about the system and the problem?

As seen, the first step in all model developments is to define the problem, and even that would often require a multidisciplinary team because it is important to quantify all the sources of the problem. The system is defined together with the problem because the problem affects either a man-made or a natural system. The problem can of course not be solved unless the processes, reactions, and the interconnections in a real system are considered. It does not imply that all details associated with the problem and the system should be considered by the model development. It is important on an early stage of the model development, however, that it is made clear for all members of the model team that the focus is the system—not the details, remembering that the system is more than the sum of its parts. On the other hand, it is important to include in the model all the important sources to the problem for the ecosystem. The selection of model complexity will be discussed in slightly more detail later in this chapter. This step in complete accordance with the steps of integrated environmental management is presented in Chapter 1.

ILLUSTRATION 9.1

Find the connectivity (number of connections/number of possible connections) of the model in Figure 9.2.

Solution

It is advantageous to calculate the connectivity of a model, which is defined as the ratio of realized connections to the number of possible connections. The conceptual diagram shows 14 connections + 5 forcing functions. The number of possible connections for seven state variables would

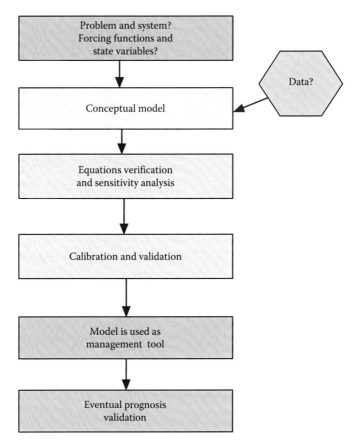

FIGURE 9.1
(**See color insert.**) A tentative modeling procedure. (Adapted from Jørgensen, S.E. and B. Fath., *Fundamentals of Ecological Modelling*, Elsevier, Amsterdam, 2011.)

be 7*7–7 = 42. It means that the connectivity is 14/42 or 0.333. It is generally accepted that the connectivity should be between 0.25 and 0.5. If it is lower than 0.25, the connections are not sufficient to ensure a good recycling of the elements, and if it is more than 0.5, the food net becomes too rigid.

It is beneficial to use a mass or energy balance to choose the most important components to be included in the model. Let us illustrate the considerations by use of an example. Let us anticipate that it is an open question, whether birds should be included in an eutrophication model. Birds contribute to the inputs of nutrients by their droppings. If the nutrients—nitrogen and phosphorus—coming from the birds' droppings are insignificant compared with the amounts of nutrients coming from drainage water, precipitation, and wastewater, inclusion of birds as a contributing component in the model is only an unnecessary complication that only would contribute to the

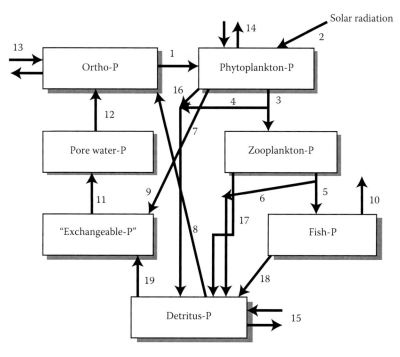

FIGURE 9.2
The phosphorus cycle. The processes are the following: (1) uptake of phosphorus by algae; (2) photosynthesis; (3) grazing with loss of undigested matter; (4) and (5) predation with loss of undigested material; (6), (7), and (9) settling of phytoplankton; (8) mineralization; (10) fishery; (11) mineralization of phosphorous organic compounds in the sediment; (12) diffusion of pore water P; (13), (14), and (15) inputs/outputs; (16), (17), and (18) mortalities; and (19) settling of detritus.

uncertainty. There are, however, a few cases where birds may contribute as much as 25% or at least more than 5% of the total inputs of nutrients. In such cases, it is of course important to include birds as a model component. A mass balance is needed to uncover the main sources of the problem—in the example the inputs of nutrients. Similarly, pesticide models mostly focus on the nonpoint pesticide pollution from agriculture and do only rarely need to include some minor point pollution sources.

In this first step, the problem and the system—for ecological models, is the system an ecosystem, several interacting ecosystems, or a part of an ecosystem?—are described by the use of words. The verbal model is, however, difficult to visualize and it is, therefore, conveniently translated in the second step of the modeling procedure into a conceptual diagram, which contains the state variables, the forcing functions, and how these components are interrelated by mathematical formulations of processes.

As shown in Figure 9.1, the construction of the conceptual diagram is possible by considering the available data. Owing to the calibration of the model

(Step 4), it is generally recommended to include only state variables that have observations. In a few cases, it may be possible to include one or two state variables that are not observed if it is difficult to get the observations and it is obvious how the nonobserved state variables are related to the other observed state variables.

Figure 9.2 illustrates a conceptual diagram of the phosphorus cycle in a lake. The next step comprises the mathematical formulation. A differential equation is formulated as mentioned above for each state variable: accumulation = inputs − outputs. The accumulation for the state variable S is dS/dt, while the inputs and outputs are processes, expressed by algebraic equations. For the conceptual diagram of the phosphorus cycle in Figure 9.2, for instance, all the shown processes should and could be expressed by an equation.

The conceptual diagram in Figure 9.2 shows the state variables as boxes and the processes as arrows between boxes, for instance, process number one. The forcing functions are symbolized by arrows to or from a state variable, and to or from the ecosystem, for instance, processes 15 and 16. It is possible to use other symbols of course for the modeling components. The software STELLA that will be used to illustrate the development of models is using boxes for state variable, thick arrows with a symbol of a valve for the processes, thick arrows coming or going to a cloud for the forcing functions (require a constant, an equation, a table, or a graph), a thin arrow to indicate the transfer of information, and a variable (forcing function, parameter and/or a state variable calculated by an algebraic expression from another state variable and so on) (see Figure 9.3).

There are other symbolic languages for the development of conceptual diagrams, for instance, Odum's energy circuit language. It has many more symbols than STELLA and is therefore more informative but of course also more time consuming to develop. For an overview of the most applied symbolic languages, including Odum's energy circuit language, see Jørgensen and Fath (2011). In this book, we will use either a simple conceptual diagram as in Figure 9.2 or a STELLA diagram (see Figure 9.3).

Figure 9.4 illustrates the idea behind the use of differential equations for the state variables. In mathematics, the differential equations are solved analytically, while the equations are solved numerically by computers. A time step is selected for the model calculations by computers. The shorter the time step, the closer will the computer calculations come to the real-time variations of inputs and outputs. It is recommended to test different time steps and use the longest time step, which would not give any significant change of the model results by decreasing the time step. Significant changes are of course evaluated relative to the accuracy of the observations that are used as the basis for the development of the model.

The software STELLA erects the differential equations from the diagram. The time derivate of the state variables will be equal to all the inputs = all process arrows going into the state variables minus all outputs = all process

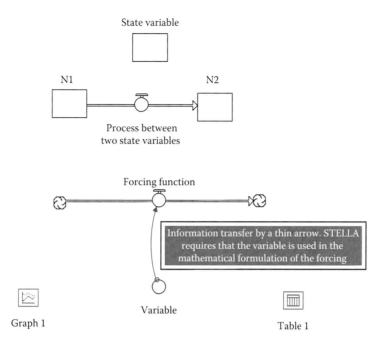

FIGURE 9.3
The symbols applied to erect a conceptual diagram using STELLA. State variables are boxes for which differential equations are erected as accumulation = inputs – outputs. Processes are thick arrows with the symbol of a valve. Forcing functions are thick arrow starting or ending as a cloud. Circles are variables in general. Graph 1 and Table 1 indicate that the results can be presented as graphs or as tables.

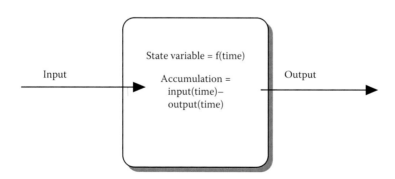

FIGURE 9.4
The idea behind the use of differential equations. The differential equation accounts for the increase of a state variable due to what is denoted in the figure accumulation for a selected time step. The accumulation is input – output. Theoretically, in mathematics, the time step is infinitely small, but when the equation is solved numerically by use of computers, we are selecting a time step. The shorter the time step, the closer is the correct mathematical solution.

arrows going out from the state variable. The processes must, however, be formulated as an algebraic equation. Table 9.1 gives an overview of the most applied process equations.

There are several modifications of these expressions. For instance, it is often used as a threshold concentration tr, in the Michaelis–Menten expression. The concentration is replaced by the concentration–tr. The concentration therefore has to exceed tr to generate any rate. For grazing and predation, it is often used to multiply the Michaelis–Menten expression by (1 – concentration/carrying capacity), similar to what is used in the logistic growth expression (see process 5 in Table 9.1). It implies that when the food is abundant (concentration is high), another factor determines the growth, for instance, the space or the nesting area.

Table 9.1 gives information about the applications of these six expressions: for which processes these equations are most often used. Most

TABLE 9.1

Process Equations

---

(1) A constant flow rate—also denoted zero-order expression.

(2) A first-order rate expression, where the rate is proportional to a variable, for instance, a concentration of a state variables: rate $= dC/dt = k*C$. This expression corresponds to exponential growth. The following expression is obtained by integration: $C(t) = C_o*e^{kt}$.

First-order decay has the rate $= dC/dt = -k*t$ and $C(t) = C_o*e^{-kt}$.

(3) A second-order rate expression, where the rate is proportional to two variables simultaneously.

(4) A Michaelis–Menten expression or Monod kinetics known from kinetics of enzymatic processes. At small concentrations of the substrate, the process rate is proportional to the substrate concentration, while the process rate is at maximum and constant at high substrate concentrations, where the enzymes are fully utilized. The same expression is used when the growth rate of plants are determined by a limiting nutrient according to Liebig's minimum law. A graph of this expression is shown in Figure 9.5. The so-called Michaelis–Menten constant or the half saturation constant corresponds to the concentration that gives half the maximum rate. At small concentrations of substrate or nutrients, the rate is very close to a first-order rate expression, while it is close to a zero-order rate expression at high concentrations. Note furthermore that the rate is regulated from a first-order to a zero-order expression more and more as the concentration increases. Note that Langmuir's adsorption isotherm that is used to describe the equilibrium of a toxic substance between soil water and soil uses the same mathematical equation.

(5) A first-order rate expression with a regulation due to limitation by another factor, for instance, the space or the nesting areas. It is expressed by introduction of a carrying capacity. The general first-order expression is applied and regulated by the following factor: (1 – concentration/carrying capacity). When the concentration reaches the carrying capacity the factor becomes zero and the growth stops. This process rate expression is denoted as logistic growth. These two growth expressions are often applied in population dynamic models. A similar equation is used to describe the microbiological decomposition of a toxic substance in water and soil and the carrying capacity is determined by the concentration of microorganisms.

(6) Rates governed by diffusion often uses a concentration gradient dC/dx to determine the rate, as it is expressed in Fick's laws: rate $= k*(dC/dx)$ (Fick's first law)

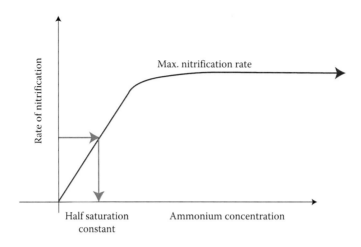

FIGURE 9.5
Graph of the Michaelis–Menten equation. In this case, the rate of nitrification (oxidation of ammonium to nitrate) versus the concentration of the substrate, ammonium, is shown. The same graph is obtained for the growth rate of phytoplankton versus the inorganic reactive phosphorus concentration, provided that phosphorus is the limiting nutrient. The figure shows that the half saturation constant is corresponding to the ammonium concentration that yields half of the maximum rate.

processes of ecological relevance are covered with these six expressions, although there are of course a few processes that will require another mathematical formulation. Most process rates are highly dependent on the temperature. Usually, the Arrhenius expression can be used to account for the influence of the temperature: rate at temperature t = rate at reference temperature (often 20°C)* $k^{(t-20)}$.

k is a coefficient varying for most processes between 1.02 and 1.12. For many biological processes, the value of growth and decomposition of toxic substances is 1.05–1.06.

The third step of the modeling procedure includes verification and sensitivity analysis. Verification is a test of the internal logic of the model. Typical questions in the verification phase are the following: Does the model react as expected? Is the model long-term stable? Does the model follow the law of mass conservation? Is the use of units consistent? It implies that all equations should be checked for a consistent use of units. Verification, to some extent, is a subjective assessment of the behavior of the model. To a large extent, the verification will go on during the use of the model before the calibration phase.

Note that during verification it is possible to perform "Gedanken experiments" similar to how Einstein tested his scientific work. We can, for instance, test an ecotoxicological model by its reactions to the following ideas: We rent a chopper and buy 1000 kg of a pesticide and drop it instantly to an ecosystem. The experiment could by use of the model be made at no costs, while it would be very expensive to rent a chopper and buy 1000 kg of pesticide. This is another clear advantage of models: It is easy to perform "Gedanken experiments."

Sensitivity analysis follows verification. Through this analysis, the modeler gets a good overview of the most sensitive components of the model. Thus, sensitivity analysis attempts to provide a measure of the sensitivity of either parameters, or forcing functions, or submodels to the state variables of greatest interest in the model. If a modeler wants to develop a model that is able to predict the toxic substance concentration in carnivorous insects because of the use of insecticides, he will obviously choose this state variable as the most important one, maybe besides the concentration of the toxic substance concentration in plants and herbivorous insects.

In practical modeling, the sensitivity analysis is carried out by changing the parameters, the forcing functions, or the submodels. The corresponding response on the selected state variable is observed. Thus, the sensitivity, S, of a parameter, P, is defined as follows:

$$S = \frac{[\partial x / x]}{[\partial P / P]} \tag{9.1}$$

where x is the state variable under consideration.

The relative change in the parameter value is chosen based on our knowledge of the certainty of the parameters. If, for instance, the modeler estimates the uncertainty to be about 50%, he will probably choose a change in the parameters at ±10% and ±50% and record the corresponding change in the state variable(s). It is often necessary to find the sensitivity at two or more levels of parameter changes as the relationship between a parameter and a state variable rarely is linear. A sensitivity analysis makes it possible to distinguish between high-leverage variables, whose values have a significant impact on the system behavior and low-leverage variables, whose values have minimal impact on the system. Obviously, the modeler must concentrate his effort on improvements of the parameters and the submodels associated with the high-leverage variables.

A sensitivity analysis of forcing functions is not in principle different from the sensitivity analysis of parameters, except that forcing functions are often indicated as table function of the time or by an equation. The table values or the equation are multiplied by factors that consider the possible uncertainty. If the uncertainty is estimated to be from, for instance, 10%–50% the factors 0.5, 0.9, 1.1, and 1.5 are used and it is recorded how much these factors would change the focal state variable. Based on these results, it is possible to determine how accurate we have to know the forcing functions as function of the time for a required certainty for the focal state variable.

A sensitivity analysis of submodels (different process equations) can also be carried out. Then the change in a state variable is recorded when the equation of a submodel is deleted from the model or changed to an alternative expression, for instance, with more details built into the submodel. Such results may be used to make structural changes in the model. If the sensitivity, for instance, shows that it is crucial for the model results to use a more detailed given submodel, this result should be used to change the model correspondingly. The selection of the complexity and the structure of the model should therefore work hand in hand with the sensitivity analysis.

The fourth modeling step encompasses calibration and validation. The scope of the calibration is to improve the parameter estimation. Some parameters in causal ecological models can be found in the literature, not necessarily as constants but as approximate values or intervals. To cover all possible parameters for all possible ecological models including ecotoxicological models, we need, however, to know more than 10 billion parameters. It is therefore obvious that in modeling there is a particular need for parameter estimation methods. A description of these methods can be found in Jørgensen and Fath (2011). Chapter 4 mentions applicable methods for the estimation of ecotoxicological parameters. Under all circumstances it is a great advantage to give even approximate values of the parameters before the calibration gets started because it is much easier to search for a value, for instance, between 1 and 4 than to search between 0 and +8.

Even where all parameters are known within intervals either from the literature or from estimation methods, it is in most cases necessary to calibrate the model. Several sets of parameters are tested by the calibration and the various model results of state variables are compared with the observed values of the same state variables. The parameter set that gives the best agreement between model results and measured values is chosen.

The need for the calibration can be explained by use of the following characteristics of ecological models and their parameters:

1. Most parameters in environmental science and ecology are not known as exact values. Therefore, all literature values for parameters (Jørgensen et al. 1991, 2000; see Figure 9.6) have an uncertainty. Parameter estimation methods must be used, when no literature value can be found. It is important particularly for ecotoxicological models.

2. All models in ecology and environmental sciences are simplifications of nature. The most important components and processes may be included, but the model structure does not account for all possible details. To a certain extent, the influence of some unimportant components and processes can be taken into account by the calibration. This will give values for the parameters slightly different from the real, but still unknown, values in nature. The difference may partly account for the influence from the omitted details.

3. By far, most models in environmental sciences and ecology are "lumped models," which implies that one parameter may represent

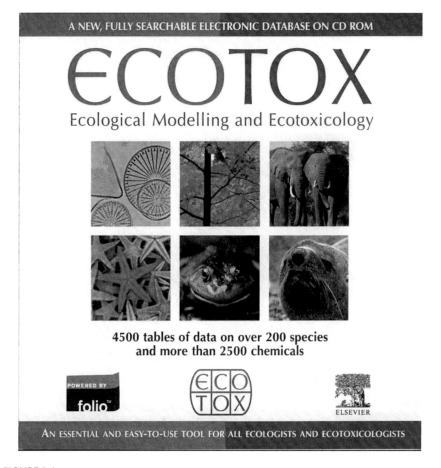

FIGURE 9.6
(**See color insert.**) One of the major literature sources for parameter. (Jørgensen, L.A. et al. 2000. Ecotox. CD.)

the average values of several species. As each species has its own characteristic parameter value, the variation in the species composition with time will inevitably give a corresponding variation in the average parameter used in the model. Adaptation and shifts in species composition will require other approaches, the so-called structurally dynamic models (SDMs).

A calibration cannot be carried out randomly if more than a couple of parameters have been selected for calibration. If, for instance, 10 parameters have to be calibrated and the uncertainties justify the testing of 10 values for each parameter, the model has to be run 10^{10} times, which of course is an impossible task. Therefore, the modeler must learn the behavior of the model by varying one or two parameters at a time and observing the response of the most crucial state variables. In some (few) cases, it is possible to separate the model into several submodels, which can be calibrated approximately independently. Although the calibration described is based to some extent on a systematic approach, it is still a trial-and-error procedure. However, procedures for automatic calibration are available. This does not mean that the trial-and-error calibration described above is redundant. If the automatic calibration should give satisfactory results within a certain frame of time, it is necessary to calibrate only 4–8 parameters simultaneously. In any circumstances, it will become easier to find the optimum parameter set, the smaller the uncertainties of the parameters are, before the calibration gets started.

In the trial-and-error calibration, the modeler has to set up, somewhat intuitively, calibration criteria. For instance, you may want to simulate rather accurately the maximum concentration of a toxic substance in a stream and/or the time at which the maximum occurs. When you are satisfied with these model results, you may then want to simulate the shape of the toxic substance concentration versus time curve properly, and so on. You calibrate the model step by step to achieve these objectives step by step.

If an automatic calibration procedure is applied, it is necessary to formulate objective criteria for the calibration. A possible criteria could be to find the parameter set that yields smallest standard deviation between modeled and observed values of all the state variables or a number of selected state variables. Often, the modeler is most interested in a good agreement between model output and observations for one or two state variables, which are considered the main model results, while he is less interested in a good agreement with the other state variables. He may furthermore choose weights for the various state variables to account for the emphasis he puts on each state variable in the model. For a model of the fate and effect of an insecticide, he may emphasize on the toxic substance concentration of the carnivorous insects and may consider the toxic substance concentrations in plants, herbivorous insects, and soil to be of less importance. He may, therefore, choose a weight of 10 for the first state variable and only one or two for the subsequent three.

If it is impossible to calibrate a model properly, it is not necessarily due to an incorrect model, but may be due to poor quality of data. The quality of the data is crucial for calibration. It is, furthermore, of great importance that the observations reflect the real dynamics of the system. If the objective of the model is to give a good description of one or a few state variables, it is of course particularly essential that the data can show the dynamics of just these internal variables. The frequency of the data collection should therefore reflect the dynamics of the focal state variables. This rule has unfortunately often been violated in modeling.

It is strongly recommended that the dynamics of all state variables are considered before the data collection program is determined in detail. Frequently, some state variables have particularly pronounced dynamics in specific periods—often in spring—and it may be of great advantage to have a dense data collection in this period, in particular. Jørgensen et al. (1981)

and Jørgensen and Fath (2011) show how a dense data collection program in a certain period can be applied to provide additional certainty for the determination of some important parameters.

From these considerations, recommendations can now be drawn up about the feasibility of carrying out a calibration of a model in ecology:

1. Find as many parameters as possible from the literature (see Jørgensen et al., 1991, 2000). Even a wide range for the parameters should be considered very valuable, as approximate initial guesses for all parameters are urgently needed.

2. If some parameters cannot be found in the literature, which is often the case, estimation methods should be used. For crucial parameters, it may be better to determine them by experiments *in situ* or in the laboratory.

3. A sensitivity analysis should be carried out to determine which parameters are most important to be known with high certainty.

4. The use of an intensive data collection program for the most important state variables should be considered to provide a better estimation for the most crucial parameters. It is generally not possible to uncover the dynamics of state variables by observations with too little frequency that do not reflect the dynamics of the system. For instance, if a model considers the diurnal variation of photosynthesis and respiration for plants, or the diurnal variations are of utmost importance for the core problem of the model, it is absolutely necessary to build the model on the basis of observations several times per 24 h. Further detail about the method see Jørgensen and Fath (2011).

9.5 Overview of Available Ecological Models

It is feasible in the modeling literature to find models of the following ecosystems:

1. Lakes
2. Estuaries
3. Rivers
4. Coastal zones
5. Coastal lagoons
6. Open sea ecosystems
7. Grasslands

8. Savannas

9. Forests

10. Polar ecosystems

11. Mountain ecosystems

12. Coral reefs

13. Wetlands (various types of wetlands, wet meadows, bogs, swamps, forested wetlands, marshes, and flood plains)

14. Deserts

15. Agricultural systems

16. Aquacultures

17. Wastewater systems

It is furthermore possible to find models focusing on the following environmental problems:

1. Oxygen depletion

2. Eutrophication

3. Acidification

4. Pollution by toxic organic compounds, including pesticides, pharmaceuticals, and endocrine disruptors

5. Pollution by heavy metals

6. Control of fishery

7. Pollution of groundwater

8. Planning of landscapes

9. Global warming and climate changes

10. Decomposition of the ozone layer and its effects

11. Groundwater pollution

12. Spreading of fire

13. Air pollution

It is possible in the modeling literature to find papers dealing with almost all combinations of these $13 \times 17 = 221$ types of models, although it will be hard to find combinations of pollution problems and polar ecosystems or mountain ecosystems, at least for mountain ecosystems above the timberline. Furthermore, models of the decomposition of the ozone layer mainly the chain of chemical processes in the ozone layer have been developed. It implies that if we classify models in accordance to the ecosystem *and* the pollution problem they attempt to solve, we will have approximately 200 classes of models—slightly less than the above-mentioned 221 possibilities.

In addition, a number of ecological models have been developed to answer ecological scientific questions as, for instance, how important is this and this process for the resistance (buffer capacity) of the considered ecosystem or what is the result of this and this chain of processes? By a statistic examination of the journal *Ecological Modelling* 2012–2013, it was found that about 15% of the models published in the journal are papers focusing on a better understanding of the importance of components and processes and forcing function for the reaction of ecosystems.

We can divide the ecosystem models, 1–17, into five groups according to how many different models is possible to find in the literature and not counting minor modifications as a change of model:

I. The ecosystems that have been modeled heavily and where it is possible to find hundreds of different models in the model literature: rivers, lakes, forests, and agricultural systems.

II. Ecosystems for which it is possible to find in the order of hundreds of different models: estuaries, wetlands, and grassland

III. Ecosystems that have been modeled many times but still less than hundred different models have been developed: wastewater systems and aquaculture systems

IV. Ecosystems that have been modeled more than 10 times but less than 25–30 times: coral reef

V. Ecosystem that have only modeled a couple of times up to a handful of times: polar ecosystems, savanna, and mountain ecosystems above the timberline

Similarly, we can divide the models of pollution problems into five groups:

I. Heavily modeled pollution problems: oxygen depletion, eutrophication, organic toxic substances, including pharmaceuticals in the environment, air pollution problems, and global warming included impacts of climate changes

II. Environmental problems that have been modeled approximately hundreds of times: heavy metal problems, acidification, and groundwater pollution

III. Environmental problems that have been modeled many times but still less than hundred different models have been developed: fire spreading and overfishing

IV. Environmental problems that have been modeled more than 10 times but less than 25–30 times: endocrine disruptors

V. Environmental problems for which it is not possible to find more than at the most a couple of models or a handful of models in the model literature: application of GMO (genetically modified organism)

9.6 Examples of Application of Models in Ecotoxicology and Environmental Chemistry

Chapter 6 has already presented ecotoxicological models as conceptual diagrams, namely, Figures 6.1, 6.2, 6.5, 6.6, and 6.7. The last three figures have numerical indications of processes. Figure 2.3 illustrates furthermore the use of fugacity models, which gives as results the distribution of a toxic substance in the spheres. In this section, further examples of models applied in ecotoxicology and environmental chemistry will be presented to illustrate this type of models.

Toxic substance models are mostly biogeochemical models because they attempt to describe the mass flows of the considered toxic substances, although there are models of the population dynamics, which include the influence of toxic substances on the birth rate and/or the mortality, and therefore should be considered as toxic substance models.

Toxic substance models differ from other ecological models in that

1. The need for parameters to cover all possible toxic substance models is great, and general estimation methods are therefore widely used. Chapter 4 is devoted to this question.

2. The safety margin, assessment factors, should be high, for instance, expressed as the ratio between the predicted concentration and the concentration that gives undesired effects (see also the discussion in Chapter 8).

3. They require possible inclusion of an effect component, which relates the output concentration to its effect. It is easy to include an effect component in the model; it is, however, often a problem to find a well-examined relationship to base it on.

4. The models need to be simple due to points 1 and 2, and our limited knowledge of process details, parameters, sublethal effects, and antagonistic and synergistic effects.

To summarize, ecotoxicological models differ from ecological models in general by

1. Often being more simple conceptually
2. Requiring more parameters
3. A wider use of parameter estimation methods
4. A possible inclusion of an effect component

Ecotoxicological models may be divided into five classes according to their structure. The five classes illustrate also the possibilities of simplification,

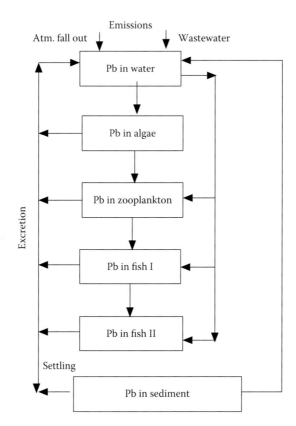

FIGURE 9.7
Conceptual diagram of the bioaccumulation of lead through a food chain in an aquatic ecosystem.

which is urgently needed as already discussed. The five classes are presented below: I–V.

9.6.1 Food Chain or Food Web Dynamic Models

This class of ecotoxicological models considers the flow of toxic substances through the food chain or food web. It can also be described as an ecosystem model focusing on the transfer of a toxic substance to ecological and nonecological components. Such models will be relatively complex and contain many state variables. The models will contain many parameters, which often have to be estimated by one of the methods presented in Chapter 4. This model type will typically be used when many organisms are affected by the toxic substance, or the entire structure of the ecosystem is threatened by the presence of the toxic substance. Because of the complexity of these models, they have not been used widely. They are similar to the more complex eutrophication models that consider the nutrient flow through the food chain or even through the food web. Sometimes they are even constructed as submodels of a eutrophication model (see, for instance, Thomann, 1984 and Thomann et al., 1974). Figure 9.7 shows a conceptual diagram of an ecotoxicological food chain model for lead. There is a flow of lead from atmospheric fallout and wastewater to an aquatic ecosystem, where it is concentrated through the food chain—by "bioaccumulation." A simplification is hardly possible for this model type because it is the aim of the model to describe and quantify the bioaccumulation through the food chain.

9.6.2 Steady-State Models of Toxic Substance Mass Flows

If the seasonal changes are minor, or of minor importance, then a steady-state model of the mass flows will often be sufficient to describe the situation and even to show the expected changes if the input of toxic substances are reduced or increased. This model type is based upon a mass balance as clearly seen from the example in Figure 6.5. It will often, but not necessarily, contain more trophic levels, but the modeler is frequently concerned with the flow of the toxic substance through the food chain. The example in Figure 6.5 considers only one trophic level.

9.6.3 Dynamic Model of a Toxic Substance in One Trophic Level

It is often only the toxic substance concentration in one trophic level that is of concern. This includes the abiotic environment (sometimes called the zeroth trophic level)—soil, water, or air. Figure 9.8 gives an example. The main concern is the DDT concentration in fish, where there may be such high concentration of DDT that, according to the WHO standards, they are not recommended for human consumption. The model can be simplified by not including the entire food chain but only the fish. Some physical–chemical reactions in the water phase are still important and they are included as shown on the conceptual diagram (Figure 9.8). As seen from these examples, simplifications are often feasible when the problem is well defined, including which component is the most sensitive to toxic matter, and which processes are the most important processes for concentration changes. The general processes of interest for modeling the concentration of a toxic substance at one trophic level are illustrated in Figure 9.9.

9.6.4 Ecotoxicological Models in Population Dynamics

Population models are biodemographic models and have the number of individuals or species as state variables. Simple population models consider only one population. Population growth is a result of the difference between natality and mortality:

$$\frac{dN}{dt} = B * N - M * N = r * N \qquad (9.2)$$

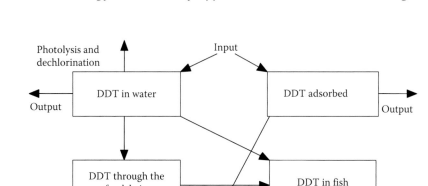

FIGURE 9.8
Conceptual diagram of a simple DDT model.

where N is the number of individuals, B is the natality, that is, the number of new individuals per unit of time and per unit of population, M is the mortality, that is, the number of organisms that die per unit of time and per unit of population; and r is the increase in the number of organisms per unit of time and per unit of population, and is equal to B–M. B, N, and r are not necessarily constants as in the exponential growth equation, but are dependent

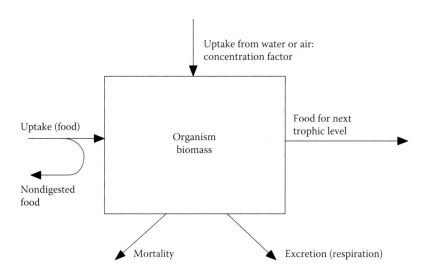

FIGURE 9.9
Processes of interest for modeling the concentration of a toxic substance at one trophic level.

on N, the carrying capacity, and other factors. The concentration of a toxic substance in the environment or in the organisms may influence the natality and the mortality, and if the relation between a toxic substance concentration and these population dynamic parameters is included in the model, it becomes an ecotoxicological model of population dynamics.

Population dynamic models may include two or more trophic levels and ecotoxicological models will include the influence of the toxic substance concentration on natality, mortality, and interactions between these populations. In other words, an ecotoxicological model of population dynamics is a general model of population dynamics with the inclusion of relations between toxic substance concentrations and some important model parameters.

9.6.5 Ecotoxicological Models with Effect Components

Although the already-presented models may include relations between concentrations of toxic substances and their effects, these are limited to, for instance, population dynamic parameters, not to a final assessment of the overall effect. In comparison, models with effect include more comprehensive relations between toxic substance concentrations and effects. These models may include not only lethal and/or sublethal effects but also effects on biochemical reactions or on the enzyme system. The effects may be considered on various levels of the biological hierarchy from the cells to the ecosystems.

In many problems, it may be necessary to go into more details of the effects to answer the following relevant questions:

1. Does the toxic substance accumulate in the organism?
2. What will the long-term concentration in the organism be when uptake rate, excretion rate, and biochemical decomposition rate are considered?
3. What is the chronic effect of this concentration?
4. Does the toxic substance accumulate in one or more organs?
5. What is the transfer between various parts of the organism?
6. Will decomposition products eventually cause additional effects and which effects do they have?

A detailed answer to these questions may require a model of the processes that take place in the organism, and a translation of concentrations in various parts of the organism into effects. This implies that the intake = (uptake by the organism)*(efficiency of uptake). Intake may either be from water or air, which also may be expressed (at steady state) by concentration factors, which are the ratios between the concentration in the organism and in the air or water. But, if all the above-mentioned processes were taken into considerations for just a few organisms, the model would easily become too complex, contain too many parameters to calibrate, and require more detailed knowledge than is possible to provide in most cases. Often, we even do not have all the relations needed for a detailed model, as toxicology and ecotoxicology are not completely well understood. Therefore, most models in this class will not consider too many details of the partition of the toxic substances in organisms and their corresponding effects, but rather be limited to the simple accumulation in the organisms and their effects. Usually, accumulation is rather easy to model and the following simple equation is often sufficiently accurate:

$$\frac{dC}{dt} = \frac{ef * Cf * F + em * Cm * V}{W} - Ex * C = \frac{INT}{W} - Ex * C \qquad (9.3)$$

where C is the concentration of the toxic substance in the organism; ef and em are the efficiencies for the uptake from the food and medium, respectively (water or air); Cf and Cm are the concentrations of the toxic substance in the food and medium, respectively; F is the amount of food uptake per day; V is the volume of water or air taken up per day; W is the body weight either as dry or wet matter; and Ex is the excretion coefficient (1/day). As can be seen from the equation, INT covers the total intake of toxic substance per day.

This equation has a numerical solution, and the corresponding plot is shown in Figure 9.10:

$$\frac{C}{C(max)} = \frac{(INT * (1 - \exp(Ex * t)))}{W * Ex} \qquad (9.4)$$

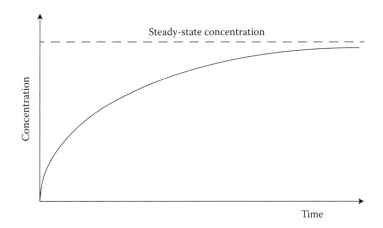

FIGURE 9.10
Concentration of a toxic substance in an organism versus time.

where C(max) is the steady-state value of C:

$$C(max) = \frac{INT}{W * Ex} \qquad (9.5)$$

Synergistic and antagonistic effects have not been touched upon so far. They are rarely considered in this type of model for the simple reason that we do not have much knowledge about these effects. If we have to model combined effects of two or more toxic substances, then we can only assume additive effects, unless we can provide empirical relationships for the combined effect.

A number of toxic substance models have been published during the last 40 years and several models are available in ecotoxicology today. During the last 10 years, many of the models developed earlier have been applied in environmental management. Most of the models reflect the proposition that good knowledge of the problem and ecosystem can be used to make reasonable simplifications. Ecotoxicological modeling has been approached from two sides: population dynamics and biogeochemical flow analysis. As the second approach has been most in focus in environmental management, it is also natural to approach the toxic substance problems from this angle. The most difficult part of modeling the effect and distribution of toxic substances is to obtain the relevant knowledge about the behavior of the toxic substances in the environment (see Chapter 7), and to use this knowledge to make the feasible simplifications. It gives the modeler of ecotoxicological problems a particular challenge by selection of the right and balanced complexity, and there are many examples of rather simple ecotoxicological models, which can solve the focal problem.

Many ecotoxicological models have been developed (see, for instance, Jørgensen and Fath, 2011, Jørgensen, 2011, and Jørgensen et al., 1996). Several accidental releases of toxic substances into the environmental have reinforced the need for models. The result has been that several ecotoxicological models have been developed in the period since the late 1970s. All the above-mentioned five types of ecotoxicological models I–V have been developed during the last 35–40 years.

9.7 Models as a Strong Management Tool: Problems and Possibilities

A management problem to be solved often can be formulated as follows: If certain forcing functions (management actions) are varied, what will their influence on the state of an ecosystem be? The model is used to answer this question or, in other words, to predict what will change in a system when the control functions managed by humans are varied over time and space.

Typical *control functions* are the consumption of fossil fuel, regulation of water level in a river by a dam, discharge of toxic pollutants, or fisheries policy.

It is important that, to a certain extent, the manager should take part in the entire development of a management model, since he will ultimately define the modeling objectives and select the modeling scenarios. The success of the application of a management model, to a large degree, is dependent on an open dialog between the modeler and the manager.

A further complexity is the construction of ecological–economic models. As we gain more experience in constructing ecological and economic models, more and more of them will be developed. It often is feasible to find a relation between a control function and economic parameters. If a lake, for instance, is a major water resource, an improvement in its water quality will inevitably result in a reduction in the treatment costs of drinking water if the same water quality is to be provided. It is also possible sometimes to relate the value of a recreational area to the number of visitors, and to how much money they spend on average in the area. In many cases, however, it is difficult to assess a relationship between the economy and the state of an ecosystem. For example, how can we assess the economic advantages of an increased transparency in a water body or the increased visibility due to reduced air pollution? Ecological–economic models are useful in some cases, but should be used with caution, and the relations between the economy and environmental conditions critically evaluated, before the model results are applied.

Data collection is the most expensive step of the model construction. For many environmental management models, it has been found that needed data collection comprises 80%–90% of the total model costs (including the costs of data collection)—often about 10 times as much as the cost of developing the model. Because complex models require much more data than simple ones, the selection of the complexity of environmental management models should be closely related to the costs involved in the environmental problem to be solved. Thus, it is not surprising that development of the most complex environmental management models have generally been limited to large ecosystems, where the economic involvement is great.

The predictive capability of environmental models can always be improved in a specific case by expansion of the data collection program, and by a correspondingly increased model complexity, provided the modelers are sufficiently skilled to know in which direction further expansion of the entire program must develop in order to improve the model's predictive capabilities.

The relation between the economy of the project and the accuracy of the model is presented in the form shown in Figure 9.11. The reduction in the discrepancy between model predictions and reality is lower for the next dollar invested in the project, because the log (cost) versus difference between model results and observations gives a straight line. But it is also clear from the shape of the curve that the associated errors can hardly be completely eliminated. All model predictions have a standard deviation associated with them. This

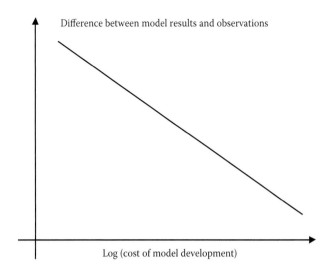

FIGURE 9.11
The more a modeler invests in a model and in data collection, the closer he/she will come to realistic predictions. However, the modeler will always gain less for the next dollar invested, and will never be able to give completely accurate predictions. With good approximations, log (costs) versus the difference between model results and observations is a straight line, corresponding to an exponentially decreasing difference with an increasing investment in the model development.

fact is not surprising to scientists, but it often is not understood or appreciated by decision makers to whom the modeler typically presents his or her results.

Engineers use safety factors to assure that a building or a bridge will last for a certain period of time, with a very low probability of breakdown, even under extreme conditions. No reputable engineer would propose using a smaller, or no, safety factor to save some concrete and reduce the costs. The reason is obvious: nobody would want to take the responsibility for even the smallest probability of a building or bridge to collapse.

When decision makers are going to make decisions on environmental issues, the situation is strangely different. Decision makers in this situation want to use the standard deviation to save money, rather than assuring a high environmental quality under all circumstances. It is the modeler's duty, therefore, to carefully explain to the decision maker all the consequences of the various decision possibilities. A standard deviation of a prognosis for an environmental management model can, however, not always be translated into a probability, because we do not know the probability distribution. It might be none of the common distribution functions, but it is possible to use the standard deviation qualitatively or semiquantitatively, translating by the meaning of the model results by the use of words. Civil engineers are more or less in the same situation, and have been successful in the past convincing decision makers of appropriate steps to be taken in various situations. There is no reason that environmental modelers cannot do the same.

However, and fortunately when the human health is threatened by toxic substances in the environment, there is a tendency to accept that the standard deviation should be used to the benefit of the human health and not the economy. For models focusing on the contamination of ecosystems, it ought to be the same practice, because toxic substances often damage ecosystems and thereby the ecosystem services, which implies that it can often be much more expensive to use the standard deviation to the benefit in the first hand of the economy, because the cost of reduced ecosystem services may exceed the cost of using more severe standard to the benefit of the environment.

It is often advantageous to attack an environmental problem in the first place with the use of simple models. They require much fewer data, and can give the modeler and decision maker some preliminary results. If the modeling project is stopped at this stage for one or another reason, a simple model is still better than no model at all, because it will at least give a survey of the problem.

Simple models, therefore, are good starting points for the construction of more complex models. In many cases, the construction of a model is carried out as an iterative process, and a step-wise development of a complex structurally dynamic model may be the result. As previously mentioned, the first step is the development of a conceptual model. It is used to get a survey of the processes and state variables in the ecosystem of concern. The next step is the development of a simple calibrated and validated model. It is used to establish a data collection program for a more comprehensive effort closer to the final selected version. However, the third model will often reveal specific model weaknesses, the elimination of which is the goal of the fourth version of the model. At first glance, this seems to be a very cumbersome procedure. However, because data collection is the most expensive part of modeling, constructing a preliminary model for optimization of the data collection program will ultimately require fewer financial resources.

A first, simple mass balance scheme is recommended for biogeochemical models. The mass balance will indicate what possibilities exist for reducing or increasing the concentration of a chemical or pollutant, which is a crucial issue for environmental management.

Point sources of pollution are usually easier to control than anthropogenic nonpoint sources, which, in turn, are more easily controlled than natural pollutant sources. Distinction can be made between local, regional, and global pollutant sources. Because the mass balance indicates the relative quantities from each source, it is possible to identify which sources should receive the initial attention (e.g., if a nonpoint regional source of pollutants is dominant, it would be pointless to concentration first on eliminating small, local point sources, unless the latter also might have some political influence on regional decisions).

It has furthermore to be recognized that the modeler and the decision maker should communicate with each other. It is recommended, in fact, that the decision maker be invited to follow the model construction process from its very first phases, in order to become acquainted with the model strength and limitations. It also is important that the modeler and the decision maker together formulate the model objectives and interpret the model results. Moreover, they should work together in all phases of the modeling exercise. Having the modeler first build a model, and then transfer it to a decision maker accompanied by a small report on the model, is not recommended.

Communication between the decision maker and the modeler can be facilitated in many ways, and it often is the primary responsibility of the modeler to do so. If a model is built as a menu system, it might be possible to teach the decision maker how to use the model in only a few hours, thereby also increasing his or her understanding of the model and its results. If an interactive approach is applied, it is possible for the decision maker to visualize a wide range of possible decisions. The effect of this approach is increased by the use of various graphic methods to illustrate the best possible decision in regard to what happens with the use of various management strategies. Under all circumstances, it is recommended that time be invested in developing a good graphic presentation of the model results to a decision maker. Even if he or she has been currently informed about a model project through all its phases, the decision maker will not necessarily understand the background and assumptions of all the model components. Thus, it is important that the model results, including the main assumptions, shortcomings, and standard deviations underlying them, are carefully presented.

References

Jørgensen, L.A., S.E. Jørgensen, and S.N. Nielsen. 2000. Ecotox. CD.

Jørgensen, S.E. (ed.). 2011. *Handbook of Ecological Models Used in Ecosystem and Environmental Management*. CRC Press, Boca Raton, 620pp.

Jørgensen, S.E. and B. Fath. 2011. *Fundamentals of Ecological Modelling*. 4th edition. Elsevier, Amsterdam, 400pp.

Jørgensen, S.E., L.A. Jørgensen, L. Kamp Nielsen, and H.F. Mejer. 1981. Parameter estimation in eutrophication models. *Ecol. Model.* 13: 111–129.

Jørgensen, S.E., S.N. Nielsen, and L.A. Jørgensen. 1991. *Handbook of Ecological Parameters and Ecotoxicology*, Elsevier, Amsterdam. Published as CD under the name ECOTOX, with L.A. Jørgensen as first editor in year 2000.

Jørgensen, S.E., B.H. Sørensen, and S.N. Nielsen. 1996. *Handbook of Environmental and Ecological Modelling*. CRC Press, Boca Raton, 672pp.

Palmeri, L., A. Barausse, and S.E. Jørgensen. 2013. *Ecological Processes Handbook*, CRC Press, Boca Raton, 386pp.

Thomann, R.V. 1984. Physico-chemical and ecological modelling the fate of toxic substances in natural water systems. *Ecol. Model.* 22: 145–170.

Thomann, R.V. et al. 1974. A food chain model of cadmium in western Lake Erie. *Water Res.* 8: 841–851.

10

Ecological Indicators and Ecosystem Services as Diagnostic Tools in Ecotoxicology and Environmental Chemistry

10.1 Role of Ecosystem Health Assessment in Environmental Management

The idea of applying an assessment of ecosystem health in environmental management emerged in the late 1980s. The parallel to the assessment of human health is very obvious. We go to our doctor for a diagnosis (What is wrong? What causes me to not feel completely healthy?) and hopefully initiate a treatment to bring us back to normal (=healthy conditions). Your doctor will apply several indicators/examinations (pulse, blood pressure, sugar in the blood and urine, etc.) before he comes up with a diagnosis and a proper treatment. The idea behind the assessment of ecosystem health is similar; see Figure 10.1. We observe that an ecosystem is not healthy and want a diagnosis: What is wrong? What has caused this unhealthy condition? And what can we do to bring the ecosystem back to normal? To answer these questions and also to follow the results of the "cure," ecological indicators are frequently applied.

Since ecosystem health assessment (EHA) emerged in the late 1980s, numerous attempts have been made to use the idea in practice and again and again environmental managers and ecologists have asked the question: which ecological indicators should we apply? It is clear today that it is not possible to find one indicator or even a few indicators that can be used generally, as some naively thought when EHA was introduced. Of course, there are general ecological indicators that are almost used every time we have to assess ecosystem health, but they are never sufficient to present a complete diagnosis—the general indicators have always to be supplemented by other indicators. Our doctor has also general indicators. He will always take your pulse, temperature, and blood pressure—they are very good general indicators, but he also always has to supplement these general indicators with other indicators that he selects according to the description of the diseases given by

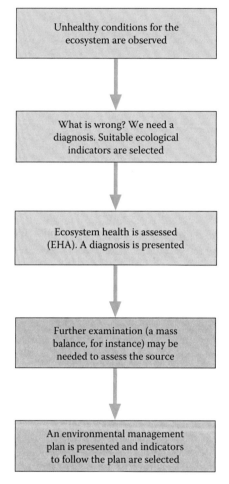

FIGURE 10.1
Illustration of how ecological indicators are used for EHA and to follow the effect of the environmental management plan.

the patient. The same process applies to the ecological doctor. If he observes dead fish but clear water, he will suspect the presence of a toxic substance in the ecosystem, while he will associate dead fish and very muddy water with oxygen depletion. In these two cases, he will use two different sets of indicators, although some general indicators may be used in both cases.

The first international conference on the application of ecological indicators for the assessment of ecosystem health was held in Fort Lauderdale, Florida, in October 1990. Since then, there have been several international and national conferences on ecological indicators and on EHA. A book titled *Ecosystem Health* edited by Costanza, Norton, and

Haskell was published in 1992 by Island Press. Blackwell published a book with the same title in 1998, edited by Rapport, Costanza, Epstein, Gaudet, and Levins. Blackwell launched a journal with the title *Ecosystem Health* in the mid-1990s with Rapport as the editor in chief. Elsevier launched a journal with the title *Ecological Indicators* in the year 2000 now with Felix Mueller as editors in chief. So, as it can be seen from this short overview of the development of the use of EHA and ecological indicators to perform the EHA, there has been a significant interest for EHA and ecological indicators.

Some may have expected that EHA would replace ecological modeling to a certain extent as it was a new method to quantify the "disease" of an ecosystem. It is also possible to assess ecosystem health based on indicators derived from observations only; on the other hand, indicators can hardly be used to make prognosis and does not provide an overview of the ecological components and their interactions as a model does. EHA and ecological modeling are rather two different and complementary tools that together give a better image of the environmental management possibilities than if either were used independently. Today, models are used increasingly as a tool to perform an EHA. The models are, furthermore, used to provide prognosis of the development of selected ecological indicators when a well-defined environmental management plan is followed.

A number of ecological indicators have been applied during the last 20 years to assess ecosystem health. As already stressed, general ecological indicators that can be applied in all cases do not exist and have at least not been found yet. A review of the literature published in the last 20 years about EHA and selection of ecological indicators reveals that there are holistic indicators that can be applied more generally and there are problem and ecosystem specific indicators that will be used repeatedly for the same problems or the same type of ecosystems. All ecosystems are different, even ecosystems of the same type are very different, there are therefore always some very case-specific indicators that are selected on the basis of sound theoretical considerations of the combination of the problem and the ecosystem. Our knowledge about human health is much more developed than our knowledge about ecosystem health, and still, there is no general procedure on how to assess a diagnosis for each of the several hundred different, possible cases a doctor will meet in his practice. We will, however, attempt in the next section to give an overview of the most applied ecological indicators for different ecosystems and their classification. It is possible to give such an overview—but not to give a general applicable procedure with a generally valid list of indicators. This does not mean that we have nothing to learn from case studies. Because the selection of indicators is difficult and vary from case to case, it is of course possible to expand one's experience by learning about as many case study as possible.

10.2 Criteria for the Selection of Ecological Indicators for EHA

Evolution of complex ecosystems can be described in terms of four major attributes (Von Bertalanffy, 1952):

1. Progressive integration (entails the development of integrative linkages between different species of biota and between biota, habitat, and climate)

2. Progressive differentiation (progressive specialization as systems evolve biotic diversity to take advantage of abilities to partition resources more finely and so forth)

3. Progressive mechanization (covers the growing number of feedbacks and regulation mechanisms)

4. Progressive centralization (it probably does not refer to a centralization in the political meaning, as ecosystems are characterized by short and fast feedbacks and decentralized control, but to the greater developed cooperation among the organisms [the Gaia effect] and the growing adaptation to all other components in the ecosystem)

Costanza et al. (1992) summarizes the concept definition of ecosystem health as follows: health as (1) homeostasis, (2) absence of disease, (3) diversity or complexity, (4) stability or resilience, (5) vigor or scope for growth, and (6) balance between system components. He emphasizes that it is necessary to consider all or least most of the definitions simultaneously. Consequently, he proposes an overall system health index, $HI = V*O*R$, where V is system vigor, O is the system organization index, and R is the resilience index. Costanza touches probably with this proposal on the most crucial ecosystem properties to cover ecosystem health.

Kay and Schneider (1992) use the term "ecosystem integrity" to refer to its ability to maintain its organization. Measures of integrity should therefore reflect the two aspects of the organizational state of an ecosystem: functional and structural. Function refers to the overall activities of the ecosystem. Structure refers to the interconnection between the components of the system. Measures of function would indicate the amount of energy being captured by the system. It could be covered by measuring the exergy captured by the system. Measures of structure would indicate the way in which energy is moving through the system. The exergy stored in the ecosystem could be a reasonable indicator of the structure.

For the use of ecological indicators for EHA from a practical environmental management point of view, seven criteria could be proposed:

1. Simple to apply and easily understood by laymen
2. Relevant in the context
3. Scientifically justifiable
4. Quantitative
5. Acceptable in terms of costs
6. Cover all relevant and actual problems
7. Sensitive to possible changes

On the other hand, from a more scientific point of view, we may say that the characteristics defining a good ecological indicator are the following:

1. Handling easiness
2. Sensibility to small variations of environmental stress
3. Independence of reference states
4. Applicability in extensive geographical areas and in the greatest possible number of communities or ecological environments
5. A possible and reliable quantification

It is not easy to fulfill all these $7 + 5 = 12$ requirements. In fact, despite the panoply of bioindicators and ecological indicators that can be found in the literature, very often the selected indicators are more or less specific for a given stress or they are applicable only to a particular type of community or ecosystem and/or scale of observation. Rarely has their wider validity in fact been utterly proved. As it will be seen through this volume, the generality of the applied ecological indicators is only limited.

The selection should at least in the first hand encompass rather too many than too little indicators because it is more easy to see after the use of redundant indicators. All stakeholders (interested in a sustainable development of the ecosystem) should usually be involved in the selection of indicators. After the selection of a handful or more indicators, each of them should be tested for the above-mentioned seven criteria, which probably will imply that the number of selected indicators is reduced.

For the various environmental problems of freshwater ecosystems, the immediate selection of indicators is often straightforward:

1. Eutrophication of aquatic ecosystems: the transparency, which is easily understood by laymen, should be supplemented by indication of phytoplankton concentration as biomass mg/L or as chlorophyll a in mg/m^3, which is of course a direct measure of the eutrophication. Also the maximum primary production as mg $C/((m^3$ or $m^2)$ 24 h) and/or as g $C/((m^3$ or $m^2)$ year) could be very informative, although it is probably not the most understandable indicators for laymen.

Furthermore, the nutrient concentrations (particularly nitrogen and phosphorus) express the eutrophication potential. The ratio between zooplankton and phytoplankton concentrations is also in some cases relevant because a relatively high zooplankton concentration may indicate that the eutrophication is under control.

2. Toxic substances: the concentration of toxic substance in water or air, in sediment or soil, and in organisms late in the food chain. Sediment and soil have often a much higher concentration than water or air, but the relevance of these concentrations depends on the water solubility and the vapor pressure of the toxic substance. Owing to biomagnification, it is recommended to use the concentration in carnivorous fish or in birds when the concentration in water is low and therefore very uncertain.

3. Acidification of ecosystems: pH, alkalinity, and pH-buffer capacity of the water in aquatic ecosystems and of the rain water.

4. Introduction and invasion of exotic species: the concentration of the introduced species and all the species that are influenced directly by the introduced species. For instance, when the Nile Perch was introduced in Lake Victoria, it would have been relevant to follow the concentration of *Tilapia* and *Haplochromis*. When the Iberian snail invaded North Europe, it would have been relevant to follow the Iberian snail as well as the competing snails.

10.3 Classification of Ecosystem Health Indicators

The ecological indicators applied today in different context, for different ecosystems and for different problems, can be classified on eight levels from the most reductionistic to the most holistic indicators. Ecological indicators for EHA do not include indicators of the climatic conditions, which in this context are considered entirely natural conditions.

Level 1 covers the presence or absence of specific species. The best-known application of this type of indicators is saprobien system (Hynes, 1971), which classifies streams in four classes according to their pollution by organic matter causing oxygen depletion: oligosaprobic water (unpolluted or almost unpolluted), beta-mesosaprobic (slightly polluted), alfa-mesoprobic (polluted), and poly-saprobic (very polluted). This classification was originally based on observations of species that were either present or absent. This level may also be applied for toxic substances as different species have different sensitivity to different toxic substances. If the composition of species is not as expected, it is difficult to find the reason. A too high concentration of toxic substances could be suspected.

Level 2 uses the ratio between classes of organisms. A characteristic example is Nygård Algae index, though hundreds of indices have been proposed over the last decade. They are difficult to apply when the health problem is due to toxic substances, but if some organisms are particularly sensitive to the presence of a specific toxic substance, then it is of course expected that the ratio of these organisms is relatively low compared to other organisms.

Level 3 is based on concentrations of chemical compounds. Examples are assessment of the level of eutrophication on the basis of the total phosphorus concentration assuming that phosphorus is the limiting factor for eutrophication or concentrations of toxic substances. Concentrations of toxic substances in the higher trophic levels are often of particular interest because the biomagnification concentrations are relatively high and can be determined with a relatively high certainty and in addition they represent worst cases. This level is the most applied level for the EHA in case of contamination by toxic substances.

Level 4 applies concentration of entire trophic levels as indicators, for instance, the concentration of phytoplankton (as chlorophyll a or as biomass per m^3) is used as an indicator for the eutrophication of lakes. A high fish concentration has also been applied as an indicator for a good water quality or birds as an indicator for a healthy forest ecosystem. A low concentration of the higher trophic levels may indicate the presence of a toxic substance because the biomagnification often represent an ecotoxicological problem.

Level 5 uses process rates as indication, for instance, primary production determinations are used as an indicator for eutrophication either as maximum gC/m^2 day or gC/m^3 day or gC/m^2 year or gC/m^3 year. A high annual growth of trees in a forest is used as an indicator for a healthy forest ecosystem and a high annual growth of a selected population may be used as an indicator for a healthy environment. A high mortality in a population can on the other hand be used as indication of an unhealthy environment due to the presence of a toxic substance. High respiration may indicate that an aquatic ecosystem has a tendency toward oxygen depletion. Growth rates are often influenced significantly by toxic substances and therefore this level may be often applied for ecosystem health problems associated with toxic substances.

Level 6 covers composite indicators, for instance, as represented by many of E.P. Odum's attributes. Examples are biomass, respiration/biomass, respiration/production, production/biomass, and ratio primary producer/consumers. E.P. Odum uses these composite indicators to assess whether an ecosystem is at an early stage of development or a mature ecosystem.

Level 7 encompasses holistic indicators such as resistance, resilience, buffer capacity, biodiversity, and all forms of diversity. Ecosystems polluted by toxic substances have often a low biodiversity because the most sensitive organisms to the toxic substances will inevitably have a high mortality. Generally, all these indicators will decrease as a consequence of toxic substances discharge.

Level 8 indicators are thermodynamic variables, which we may call super-holistic indicators as they try to see the forest through the trees and capture the total image of the ecosystem without inclusion of details. Such indicators are exergy (work energy content, see Section 10.5 for methods to calculate the work energy of ecosystems), emergy, exergy destruction (the amount of work energy lost as heat energy), entropy production, power, mass, and/or energy system retention time. The economic indicator cost/benefit (which includes all ecological benefits—not only the economic benefits of the society) also belongs to this level. They are closely associated with the ecosystem properties and will therefore indeed be able to reflect the presence of toxic substances.

10.4 Ecosystem Services

Humankind benefits in a number of ways from ecosystems. Collectively, these benefits are defined as ecosystem services. The millennium ecosystem assessment (MEA) defines ecosystem services as the benefits people can obtain from ecosystems. Ecosystem services can be classified into four broad categories: *provisioning*, such as the production of food and water; *regulating*, such as the control of climate and disease; *supporting*, such as nutrient cycles and crop pollination; and *cultural*, such as aesthetic and recreational benefits. To help inform decision makers, many ecosystem services are being assigned economic values, as mentioned in Chapter 1, and the next section will attempt to assess the value of ecosystem services for a number of ecosystems.

The services in the four categories include the following items.

10.4.1 Provisioning Services

"Products obtained from ecosystems":

- Food (including seafood and game), crops, wild foods, and spices
- Raw materials (including lumber, skins, fuel wood, organic matter, fodder, and fertilizer)
- Genetic resources (including crop improvement genes, and health care)
- Water
- Minerals (including diatomite)
- Medicinal resources (including drugs, pharmaceuticals, chemical models, and test and assay organisms)

- Chemical resources (the number of possibly useful chemicals is enormous)
- Energy (hydropower, biomass fuels)
- Ornamental resources (including fashion, handicraft, jewelry, pets, worship, decoration, and souvenirs such as furs, feathers, ivory, orchids, butterflies, aquarium fish, and shells)

10.4.2 Regulating Services

Benefits obtained from the regulation of ecosystem processes:

- Carbon sequestration and climate regulation
- Waste decomposition and detoxification
- Purification of water and air
- Pest and disease control
- Buffering water quantities (reduce thereby the number of floods and droughts)
- Moderating general weather extremes
- Maintaining diversity
- Preserving quantities (reduce erosion) and qualities (maintain fertility) of soil

10.4.3 Supporting Services

Ecosystem services are necessary for the production of all other ecosystem services:

- Nutrient dispersal and cycling
- Seed dispersal
- Primary production
- Pollinating crops and natural vegetation

10.4.4 Cultural Services

Nonmaterial benefits people obtain from ecosystems through aesthetic enrichment and experiences, reflection, and recreation:

- Cultural (including use of nature as motif in books, film, painting, folklore, national symbols, architecture, and advertising)
- Historical (including use of nature for religious or heritage value)
- Aesthetical (enjoying nature and nature as source of inspiration in art)

- Recreational experiences (including ecotourism, outdoor sports, and recreation)
- Science and education (including use of natural systems for school excursions and scientific discovery)

All the listed services will be reduced by the presence of a toxic substance in a concentration which deteriorates the components of the ecosystem. It is often and particularly the case for regulating and supporting services.

10.5 Value of Ecosystem Services

How can we estimate the value of the ecosystem services? Costanza et al. (1997) have found the value by adding the costs of all the services that we utilize by ecosystems: purification of air and water, recycling, recreational services, and natural resources such as timber, fish, and drinking water. All activities require energy that can do work. Services by ecosystems relate to activities offered to the user of ecosystems and that could therefore be measured by the work energy. It means as the total amount of eco-exergy (work capacity), that an ecosystem offers. The sustainability of nature (ecosystems) can also be expressed by the work capacity, because

1. Sustainability in Brundtland's sense means that the same level of ecosystem services must be maintained for the future generation
2. The amount of energy to be used to break down an ecosystem is equal to the work capacity of the ecosystem (Svirezhev 1998)

This idea to measure ecosystem services and thereby sustainability by eco-exergy or work capacity is pursued in this section and the results are compared with results reported by Costanza et al. (1997). It is a newly proposed approach, which inevitably will lead to the recommendation to apply eco-exergy or work energy capacity calculations of ecosystems as informative holistic indicators to assess the economic value of ecosystem services. How do we calculate the work capacity of ecosystems?

Work energy is part of the total energy that can do work. It can therefore be found as the gradient (=difference in potential) × extensive descriptor, dependent on the energy form, for instance (Jørgensen et al., 2007; Jørgensen, 2012)

- Chemical work energy = $(\mu_1 - \mu_2)^*N$, where = $(\mu_1 - \mu_2)$ is the difference in chemical potential before and after the actual chemical process and N is the number of moles participating in the chemical process.

- Pressure work energy = (p1 – p2)(–V), where p1 – p2 is the pressure difference available for the volume V.
- Potential work energy = (h1 – h2)*m*g, where h1 and h2 are the high and low altitudes, respectively, m is the mass, and g the gravity constant.
- Electrical work energy = (V1 – V2)*Q, where V1 – V2 is the difference in voltage and Q is the charge.

We can distinguish between technological work energy, also denoted as exergy, and ecological work energy also denoted as eco-exergy. Also included, in contrast to technological work energy, is the work energy of information. Technological exergy uses the environment as reference state and is useful to find the first-class energy (work) that a power plant can produce. Eco-exergy uses as reference state the same ecosystem with the same temperature and pressure but at thermodynamic—chemical equilibrium; for details, see Jørgensen (2012).

For calculation of ecosystem services, we will use the work capacity of ecosystems or eco-exergy, which includes the work energy of the information that the organisms use to perform the life processes. As the reference system has the same temperature and pressure and the magnetic and electrical work is negligible, only the chemical work energy contributes to the work capacity of ecosystems. The following equation is used for the calculations:

$$\text{Eco-exergy} = \sum_{i=1}^{i=n} \beta_i * c_i \qquad (10.1)$$

where β is a weighting factor, considering the information carried by the organisms in their genes, that is, how many amino acids in the right sequence is required to make up the organism or how much information does an organism contain? β values found based on the genome size and by indirect methods can be seen in Table 10.1.

The value of the annual ecosystem services can be expressed by the annual increase of work capacity (content) or eco-exergy. The annual growth of work energy embodied in the biomass (see Jørgensen et al., 2000) for the most common types of ecosystems can be found in general ecology textbooks or on Internet. An average β value for various ecosystems can be estimated based on the organisms that can be found generally in various types of ecosystems. Table 10.2 shows the annual biomass growth in MJ/m^2 using an average energy (exergy) content of biomass of 18.7 kJ/g, the applied average β values and the work capacities in GJ/ha year are found as the product of the biomass growth per m^2 and the β value.

TABLE 10.1

β Values = Eco-Exergy (Work Energy Including the Work Energy of Information) Content Relative to the Eco-Exergy of Detritus

Organisms	Plants	Animals	
Detritus	1.00		
Viroids	1.0004		
Virus		1.01	
Minimal cell		5.0	
Bacteria	8.5		
Archaea	13.8		
Protists (algae)	20		
Yeast		17.8	
		33	Mesozoa, Placozoa
		39	Protozoa, amoeba
		43	Phasmida (stick insects)
Fungi, molds	61		
		76	Nemertina
	91		Cnidaria (corals, sea anemones, jellyfish)
Rhodophyta	92		
		97	Gastrotricha
Porifera, sponges	98		
		109	Brachiopoda
		120	Platyhelminthes (flatworms)
		133	Nematoda (round worms)
		133	Annelida (leeches)
		143	Gnathostomulida
Mustard weed	143		
		165	Kinorhyncha
Seedless vascular plants	158		
		163	Rotifera (wheel animals)
		164	Entoprocta
Moss	174		
		167	Insecta (beetles, flies, bees, wasps, bugs, and ants)
		191	Coleodiea (sea squirt)
		221	Lepidoptera (butterflies)
		232	Crustaceans
		246	Chordata
Rice	275		
Gymnosperms (including *Pinus*)	314		
		310	Mollusca, Bivalvia, Gastropoda
		322	Mosquito

(Continued)

TABLE 10.1 (*Continued*)

β Values = Eco-Exergy (Work Energy Including the Work Energy of Information) Content Relative to the Eco-Exergy of Detritus

Organisms	Plants	Animals	
Flowering plants	393	499	Fish
		688	Amphibia
		833	Reptilia
		980	Aves (Birds)
		2127	Mammalia
		2138	Monkeys
		2145	Anthropoid apes
		2173	*Homo sapiens*

Source: Jørgensen, S.E. et al. 2005. *Ecological Modelling* 185: 165–176.
Note: β values = eco-exergy content relatively to the eco-exergy of detritus.

The work capacity of various ecosystems can easily be compared with Costanza et al.'s (1997) values by using the energy costs:

- 1 MJ has the value of 1 €-cent or 1.4 $-cent
- 1 GJ has therefore the value of 10 € or 14 $

The values found on basis of the results in Table 10.2 are compared with Costanza et al.'s results in Table 10.3. The work capacity values give a much higher money value, which is not surprising, because the work capacities include all the possible services that the ecosystems offer—not only the

TABLE 10.2

Work Energy Capacity Used to Express the Ecosystem Services for Various Types of Ecosystems

Ecosystem	Biomass Production MJ/m² year	Information Factor β Value	Work Capacity GJ/ha year
Desert	0.9	230	2070
Open sea	3.5	68	2380
Coastal zones	7.0	69	4830
Coral reefs, estuaries	80	120	96,000
Lakes, rivers	11	85	9,350
Coniferous forests	15.4	350	53,900
Deciduous forests	26.4	380	100,000
Temperate rainforests	39.6	380	150,000
Tropical rainforests	80	370	300.000
Tundra	2.6	280	7280
Croplands	20.0	210	42,000
Grassland	7.2	250	18,000
Wetlands	18	250	45,000

Note: It is calculated as biomass*information factor.

TABLE 10.3

Value of Annual Ecosystem Services

Ecosystem	k€/ha year (Based on Eco-Exergy)	$/ha year According to Costanza et al. (1997)	Ratio
Desert	20.7	?	–
Open sea	23.8	252	94
Coastal zones	48.3	4052	12
Coral reefs, estuaries	960	14,460	66
Lakes, rivers	93.5	8500	11
Coniferous forests	539	969	556
Deciduous forests	1000	969	1032
Temperate rainforests	1580	?	–
Tropical rainforests	3000	2007	1495
Tundra	72.8	?	–
Croplands	400	92	4348
Grassland	180	232	775
Wetlands	450	14,785	30.4

Source: Jørgensen, S.E. 2010. *Ecol. Complex.* 7: 311–313.

services that we are utilizing. The ratio of the two sets of values are included in the last column of Table 10.3. We evidently use the ecosystems differently, which can explain the different ratios. Notice that both sets of values are reduced by the presence of toxic substances as the ecosystem components are influenced by toxic substances and all services and the work energy of the ecosystems are thereby reduced. The ecosystems can be divided into five classes according to the different ratios, starting with the class that is utilized mostly by man for a series of services:

- A. Coastal zones, lakes, rivers: regulation, water supply, waste treatment, recreation, genetic resources, pollination, nutrient cycles, biological control, food production, refugia, transportation, raw materials, purifying air, moderating weather extremes, cultural; ratio about 10–20

- B. Wetlands: regulation, water supply, waste treatment, recreation, raw material, genetic resources, pollination, nutrient cycles, biological control, refugia, cultural; buffering water quantities, ratio about 30

- C. Open sea, estuaries, coral reef: Only climate and gas regulation, very little waste treatment, much less recreation than A and B, raw material, genetic resources, pollination, nutrient cycles (minor) biological control, (minor) refugia, raw materials, cultural; ratio about 60–100

- D. Forests, croplands, grasslands, and deserts: Mainly as raw materials, too little genetic resources, pollination, nutrient cycles, biological control, (minor) refugia, cultural, recreation; ratio about >500

- E. Cropland: Croplands are only utilized to produce raw materials (mainly food); the ratio is therefore high, 4348

The results can be summarized in the following conclusions:

- The total value of all services offered by the ecosystems may be estimated from the work capacity (eco-exergy) of the ecosystems.
- It is also a measure of the sustainability.
- We can divide the ecosystems into five classes according to how much we are able to utilize the entire spectrum of possible services.
- The sequence of our utilization of the ecosystem services is lakes and rivers, coastal zones, wetlands, estuaries, open sea ecosystems, grasslands, forests, and croplands.
- The sequence is understandable and explainable.

The calculations of eco-exergy (work energy with information included) can be used as measure for

- The entire spectrum of all the services that the ecosystems offer— not only the services that we actually utilize
- The sustainability—at least indirectly

It is therefore possible to use the calculations of eco-exergy of ecosystems as very informative holistic ecological indicators.

References

Costanza, R., B.G. Norton, and B.D. Haskell. (eds.) 1992. *Ecosystem Health: New Goals for Environmental Management*. Island Press, Washington, DC.

Costanza, R. et al. 1997. The value of the world's ecosystem services and natural capital. *Nature* 387: 252–260.

Hynes, H.B.N. 1971. *Ecology of Running Water*. Liverpool University Press, Liverpool, England.

Jørgensen, S.E. 2010. Ecosystem services, sustainability and thermodynamic indicators. *Ecological Complexity* 7: 311–313.

Jørgensen, S.E. 2012. *Introduction to Systems Ecology*. CRC Press, Boca Raton. Chinese edition 2013, 320pp.

Jørgensen, S.E., B. Fath, S. Bastiononi, M. Marques, F. Müller, S.N. Nielsen, B.C. Patten, E. Tiezzi, and R. Ulanowicz. 2007. *A New Ecology: Systems Perspectives*. Elsevier, Amsterdam, 275pp.

Jørgensen, S.E., Ladegaard, N., Debeljak, M., and Marques, J.C. 2005. Calculations of exergy for organisms. *Ecological Modelling* 185: 165–176.

Jørgensen, S.-E., B.C. Patten, and M. Straškraba. 2000. Ecosystems emerging: 4. Growth. *Ecological Modelling* 126: 249–284.

Jørgensen, S.E., R. Costanza, and Fu-Liu Xu. 2010. *Handbook of Ecological Indicators for Assessment of Ecosystem Health.* 2nd edition, CRC Press, Boca Raton, 482pp.

Kay, J. and Schneider, E.D. 1992. Thermodynamics and measures of ecological integrity. In: *Proceedings of the Ecological Indicators.* Elsevier, Amsterdam, pp. 159–182.

Svirezhev, Yu.M. 1998. Thermodynamic orientors: How to use thermodynamic concepts in ecology? In: F. Müller and M. Leupelt (eds.) *Eco Targets, Goal Functions and Orientors.* Springer-Verlag, Berlin, pp. 102–122.

Von Bertalanffy, L. 1952. *Problems of Life.* Wiley, New York.

11

Application of Environmental Technology in Environmental Management

11.1 Introduction

Environmental technology has been used for more than 100 years to solve pollution problems. Biological treatment of wastewater has been applied although in modest scale in Europe even before the start of the twentieth century. Before Rachel Carson's book *The Silent Spring*, it was the only tool available for the solution of pollution and environmental problems in general. Environmental technology refers to treatment of wastewater, smoke, and solid waste through the use of technological methods. Environmental technology is best fitted to solve pollution problems from point sources because it makes possible the use of the technological method directly on the point source, which implies that the amount of wastewater, smoke or solid waste is limited and rather concentrated. It is therefore possible to develop, in most cases, a technological method that is able to reduce the concentration of pollutants with a high efficiency. Environmental technology requires in many cases a relatively high investment, and the operational costs are sometimes high, too, although there are also many examples of relatively low operational costs. After the development of the other tools to solve pollution problems (presented in Chapters 12 and 13 and environmental legislation, see Section 13.7 and Jørgensen et al., 2015), it is clear that environmental technology often has to be beneficially combined with the other tools to find the most effective and moderate-cost solution, although there are also several cases where environmental technology is the only obvious method to apply. This is particularly the case for industrial pollution problems, in which environmental technology may even offer possibilities to recover or recycle raw material and simultaneously to remove toxic substances.

This chapter provides a short overview of the most applied environmental technological methods, of course with emphasis on solution to pollution problems associated with toxic substances. The overview of water pollution methods is presented in Section 11.2, while Section 11.3 is devoted to air pollution treatment methods and Section 11.4 focuses on the treatment of solid waste problems.

11.2 Application of Environmental Technology to Solve Wastewater Problems

Impact on aquatic ecosystems from point sources originates from discharge of wastewater. Wastewater discharges into ecosystems are man-controlled forcing functions of crucial importance for the water quality. It is, however, possible in many situations to control it completely, either by water diversion or by wastewater treatment methods. Water diversion, however, means that another water body has to cope with the pollutant load. Thus, treating the wastewater properly should be considered a generally more acceptable solution to the problem. This gives rise to two questions, namely

1. Is it possible to solve all pertinent wastewater problems?
2. What is understood by a proper wastewater treatment?

The water pollution problems associated with municipal and industrial wastewaters include their content of

- Nutrients causing eutrophication
- Biodegradable organic matter causing oxygen depletion
- Bacteria and virus affecting the sanitary quality of water, which is of particular importance, when the water is used for bathing, swimming, and drinking purposes
- Heavy metals, namely, lead, zinc, and cadmium from gutters, heavy metals from fungicides and other agricultural chemicals, and a wide range of other heavy metals in minor concentrations
- Refractory organic matter, originating mainly from industries, hospitals, and the use of pesticides, and even from the use of a wide spectrum of household articles

Tables 11.1 and 11.2 (Jørgensen, 2000) provide an overview of a wide range of wastewater treatment methods, their efficiencies, and approximate costs. Clearly, a method with good approximations is available to virtually any of the mentioned problems, as long as point sources of pollution are considered.

Industrial wastewaters can cause the same water pollution problems as municipal wastewater plus a few more associated with toxic organic and/ or inorganic compounds (particularly heavy metals and persistent organic pollutants). However, it is often necessary to solve at the source the problems associated with industrial wastewater, which can hardly be solved with municipal wastewater treatment methods. It is also the general legislation in most of the world today that industries are obliged to treat the wastewater before discharging into public sewage system. In many countries, the practice of the "polluter pays" principle has forced the industries to solve their

TABLE 11.1

Survey of Generally Applied Wastewater Treatment Methods

Method	Pollution Problem	Efficiency	Costs (US$/100 m^3)
Mechanical treatment	Suspended matter removal	0.75–0.90	3–5
	BOD$_5$ reduction	0.20–0.35	
Biological treatment	BOD$_5$ reduction	0.70–0.95	25–40
Flocculation	Phosphorus removal	0.3–0.6	6–9
	BOD$_5$ reduction	0.4–0.6	
Chemical precipitation Al$_2$(SO$_4$)$_3$ or FeCl$_3$	Phosphorus removal	0.65–0.95	10–15
	Reduction of heavy metals concentrations	0.40–0.80	
	BOD$_5$ reduction	0.50–0.65	
Chemical precipitation Ca(OH)$_2$	Phosphorus removal	0.85–0.95	12–18
	Reduction of heavy metals concentrations	0.80–0.95	
	BOD$_5$ reduction	0.50–0.70	
Chemical precipitation and flocculation	Phosphorus removal	0.9–0.98	12–18
	BOD$_5$ reduction	0.6–0.75	
Ammonia stripping	Ammonia removal	0.70–0.95	25–40
Nitrification	Ammonium oxidized to nitrate	0.80–0.95	20–30
Active carbon adsorption	COD removal (toxic substances)	0.40–0.95	60–90
	BOD$_5$ reduction	0.40–0.70	
Denitrification after nitrification	Nitrogen removal	0.70–0.90	15–25
Ion exchange	BOD$_5$ reduction (e.g., proteins)	0.20–0.40	40–60
	Phosphorus removal	0.80–0.95	70–100
	Nitrogen removal	0.80–0.95	45–60
	Reduction of heavy metal concentration	0.8–0.99	15–30
Chemical oxidation (e.g., with Cl$_2$)	Oxidation of toxic compounds	0.90–0.98	60–100
Extraction	Heavy metals and other toxic compounds	0.50–0.95	80–120
Reverse osmosis	Removes most pollutants with high efficiency, but is expensive		100–200
Disinfection methods	Reduction of microorganisms	High, can hardly be indicated	6–10
Ozonation + active carbon adsorption	Removal of refractory compounds	0.5–0.95	100–120

TABLE 11.2

Efficiency Matrix Relating Pollution Parameters and Wastewater Treatment

	Suspended Matter	BOD$_5$	COD	Total Phosphorus	Ammonium Nitrogen
Mechanical treatment	0.75–0.90	0.20–0.35	0.20–0.35	0.05–0.10	~0
Biological treatment[a]	0.75–0.95	0.65–0.90	0.10–0.20	0.05–0.10	~0
Chemical precipitation	0.80–0.95	0.50–0.75	0.50–0.75	0.80–0.95	~0
Ammonia stripping	~0	~0	~0	~0	0.70–0.96
Nitrification	~0	~0	~0	~0	0.80–0.95
Active carbon adsorption[a]	–	0.40–0.70	0.40–0.95	~0.1	High[b]
Denitrification after nitrification	~0	–	–	~0	–
Ion exchange	–0.40	0.20–0.50	0.20–0.95	0.80–0.95	0.80–0.95
Chemical oxidation	–	Corresponding to oxidation	~0	~0	~0
Extraction	–	Corresponding extraction of toxic compounds	~0	~0	~0
Reverse osmosis[a]	High but varying by the operational conditions				
Disinfection methods	–	Much corresponding to application of chlorine, ozone, etc.			

(*Continued*)

TABLE 11.2 (*Continued*)

Efficiency Matrix Relating Pollution Parameters and Wastewater Treatment

	Total Nitrogen	Heavy Metals	*Escherichia coli*	Color	Turbidity
Mechanical treatment	0.10–0.25	0.20–0.40	–	0.80–0.98	–
Biological treatment[a]	0.10–0.25	0.30–0.65	Fair	~0	–
Chemical precipitation	0.10–0.60	0.80–0.98	Good	0.30–0.70	0.80–0.98
Ammonia stripping	0.60–0.90	~0	~0	~0	~0
Nitrification	0.80–0.95	~0	Fair	~0	~0
Active carbon adsorption[a]	High[b]	0.10–0.70	Good	0.70–0.90	0.60–0.90
Denitrification after nitrification	0.70–0.90	~0	Good	~0	–
Ion exchange	0.80–0.95	0.80–0.95	Very good	0.60–0.90	0.70–0.90
Chemical oxidation	~0	~0	~0	0.60–0.90	0.50–0.80
Extraction	~0	0.50–0.95	~0	~0	~0
Reverse osmosis[a]					
Disinfection methods	Very high	0.50–0.90	0.30–0.60		

[a] Depends on the composition.
[b] As chloramines.

pollution problems to keep the production costs low. The major portion of toxic substances is therefore removed by the industries today, at least in most industrialized countries. Previously, they would only have been partially removed, if at all, at municipal wastewater treatment plants, and/or could contaminate the sludge produced at municipal wastewater treatment plants, thereby eliminating the possibility of the use of the sludge as a soil conditioner.

Moreover, the removal of high concentrations of biodegradable organic matter at the source is strongly recommended, since it is usually much more cost effective to remove these components, at least partially, when they are present in high concentrations. High concentrations of biodegradable organic matter are found in wastewater from slaughterhouses, starch factories, fish industries, dairies, and canned food industries. The removal of x% of BOD_5) costs as a rule of thumb the same independent on the level of BOD_5. It means that 1 kg of BOD_5 can be removed at a much lower price from wastewater with a high BOD_5 as, for instance, wastewater from slaughterhouses, starch factories, fish industries, dairies, and canned food industries.

The listed methods are often used in combinations of two or more steps to obtain the overall removal efficiency required by the most cost-moderate solution. The methods can also be applied in combination with ecotechnology (Chapter 12) and cleaner technology (Chapter 13). Because wastewater treatment often is costly, it is recommendable in the planning phase to examine *all* possible combinations of treatment options in order to identify the most feasible and appropriate one.

Many existing municipal wastewater treatment plants were constructed years or decades ago, and may not meet today's higher standards. Nevertheless, upgrading existing wastewater treatment plants is possible, and may be more cost moderate than building new ones (Novotny and Somlyódy, 1995; van Loosdrecht, 1998). Because the funding allocated to pollution abatement is often limited, the overall effect of upgrading wastewater treatment plants that can be upgraded with sufficient efficiency will be to the benefit of the environment. An attractive solution is often to introduce *tertiary treatment* by chemical precipitation and flocculation in an existing mechanical–biological treatment plant, with the addition of chemicals and flocculants before the primary sedimentation phase. The installation costs for this solution are minor and the additional running costs are limited to the costs of chemicals. The result is 85%–95% removal of phosphorus at low cost and 40%–60% of most heavy metals. Similarly, nitrification and denitrification, ensuring 80%–85% removal of nitrogen, can be realized with the installation of additional capacity for biological treatment (the overall water retention time in the plant is increased by 4–12 h, depending on the standards and composition of the wastewater), which is considerably less costly than installation of a completely new treatment plant. For details, see Hahn and Muller (1995) and Henze and Ødegaard (1995).

The second question refers to the selection of the right standards for the treated wastewater. It is possible to obtain removal efficiency of any pertinent parameters (BOD_5, nutrients, bacteria, viruses, toxic organic toxic compounds, color, taste, and heavy metals) with a suitable combination of the available treatment methods. However, what removal efficiencies are needed in the focal case? Because wastewater treatment is costly, the maximum allowable concentrations should not be set significantly lower than what the ecosystem receiving the effluents can tolerate. The ban of phosphate detergents to decrease phosphorus concentrations in municipal wastewater treatment plant effluents together with increased efficiency of detergents are points to consider in this context, as the treatment costs can be reduced considerably by the introduction of phosphorus-free detergents, which is more or less the situation today in most industrialized countries. However, it might be even more expensive to install an insufficient treatment plant. Thus, the potential effects of a wide range of possible pollutant inputs on water quality and on the entire lake or reservoir should be assessed, as the basis for selecting an acceptable option. This will require a quantification of the impacts of various possible pollutant inputs, considering a wide range of solutions. All processes and components affected significantly by the impacts should be included in the quantification. It is usually very helpful to develop a water quality/ecosystem model and use it properly to assist in the selection of specific environmental treatment methods. It is important to emphasize that a model has an uncertainty in all its predictions that must be considered in making a final decision. Thus, it is essential to use safety factors to the benefit of the environment in order to ensure that the selected treatment methods will have the anticipated effects. Chapter 8 discusses the safety factors when ERA is developed and Chapter 4 discusses the uncertainty associated with the use of estimation methods. Chapter 9 discusses the use of ecological model and additionally the very relevant core question: should the uncertainty be used to the benefit of the environment or the economy? If the uncertainty is taken into account for the sake of economy, as it is unfortunately often done, the investment may be wasted because the foreseen recovery of the ecosystem will not be realized.

Application of the methods identified in Tables 11.1 and 11.2 gives only approximate results, and the indications should therefore be used with caution. However, first estimates, such as those shown in the tables, are useful for evaluations of various alternative solutions to wastewater pollution problems. The biological treatment may be either an activated sludge plant or a trickling filter.

The cost of treating 100 m^3 of wastewater is also included in Table 11.1. It is based on approximate indications because the costs vary from place to place, as the costs of labor are very different in different countries and are highly dependent on the size of the wastewater treatment plant. The costs are calculated as the running costs (electricity, labor, chemicals, and maintenance), plus 10% of the investment to cover interests and annual appreciation. The annual water consumption of one person in an industrialized country corresponds to approximately 50 maximum 100 m^3 and in average about 60 m^3.

A problem in many developing countries is the relatively high cost of wastewater treatment. Although this cost might justify diversion of wastewater, the application of "soft technology"—"ecotechnology"—should be considered, too. Some corresponding methods will be touched upon in Chapter 12, but proper planning at an early phase, and considering all predictable problems, offers the widest range of cost-effective possibilities, and may enable prevention of the pollution problems before they become serious and even irreversible.

Corrections at a later stage, when pollution has already degraded the water quality and associated ecosystems, are possible but will always be more expensive than the costs of proper wastewater treatment at an early stage. This is due, in part, to the fact that the accumulation of pollutants in an aquatic ecosystem over time will always cause additional problems and, therefore, result in additional costs. Thus, pollution prevention at an early stage is better than curing pollution at a later stage. Removal of phosphorus from wastewater at an early stage, for example, is always beneficial since the surplus phosphorus will accumulate in the sediments of aquatic ecosystems—to a large extent—and allow its remobilization back into the water column under certain chemical conditions in the water body.

Model studies reveal the time required to restore an ecosystem, or how much higher phosphorus removal efficiency will be required to compensate for each year that the implementation of an appropriate phosphorus removal technology is postponed, as phosphorus is accumulated in the ecosystem, mainly in the sediment. However, it is not unusual that implementation of a phosphorus removal technology a few years later than it was first feasible may delay the restoration of an aquatic ecosystem by one or more decades, due to the fact that the additional phosphorus accumulated in the sediments may significantly increase the quantity of phosphorus in the water column.

Important pollution sources may be reduced if the liquid waste and sludge land disposal is avoided or minimized. For municipal areas, two options are used to decrease the hydraulic load of wastewater treatment plants, including

1. Decreasing water use, thereby saving water and producing smaller volumes of polluted water
2. Separating storm water from municipal domestic waste with a similar result

One result of these options is that the capacity of wastewater treatment plants can be kept smaller, achieving significant cost savings.

The selection of proper wastewater treatment methods for point sources of pollution is summarized in the following points:

- Develop models for the impact of the wastewater on freshwater ecosystems, considering the impact on the water quality and the entire ecosystem.

- Apply the model to identify the maximum allowable pollutant concentration in the treated wastewater. Any uncertainty associated with the model predictions should be reflected in identifying the lower maximum allowable concentrations.
- Select the combination of available treatment methods able to meet the standards at the lowest costs without impacting the proper operation of the plant.
- If the investment needed for a proper solution to a problem cannot be provided, the application of cost-moderate technology that will reduce the accumulation of pollutants in the aquatic ecosystem should be considered. Any measures taken at an early stage will reduce the costs at a later stage.

The composition of domestic sewage varies surprisingly very little from place to place, although to a certain extent it reflects the economic status of the society.

Models are used increasingly to design and to optimize wastewater treatment methods. For details about the applied models, refer to Jørgensen (2011) *Handbook of Ecological Models Used in Ecosystem and Environmental Management*, which has a comprehensive chapter devoted to models of wastewater treatment systems.

Aluminum sulfate, various polyaluminates, calcium hydroxide, and iron(III) chloride can be applied as chemicals for the precipitation, which is a frequently applied step in the treatment of municipal wastewater. As previously indicated, the heavy metals will be precipitated, too, and the efficiency is dependent on the heavy metal that we consider. The removal efficiency is for most heavy metals however about 40%–60% in practice. High efficiency is, however, obtained for heavy metals, when calcium hydroxide is used as precipitant. Figure 11.1 shows the different combinations of mechanical–biological treatment of municipal wastewater and chemical precipitation.

Precipitation can furthermore be applied on industrial wastewater containing heavy metals and in that case calcium hydroxide should be applied as a precipitant with the heavy metal hydroxides to have a low solubility. In most cases, organic toxic substances can be removed by precipitation, while the efficiency is dependent on the chemical structure of the compound.

The other methods applicable to remove heavy metals and toxic organic compounds are as indicated in Table 11.1: ion exchange, chemical oxidation, and adsorption by activated carbon. Ion exchange can of course only be applied to remove ionic compounds, which means that it can remove heavy metal ions effectively. It is possible in some cases to select an ion exchanger that has a particular high removal selectivity and efficiency for a specific heavy metal. Oxidation by use of, for instance, chlorine or ozone can be used to decompose toxic organic compounds. Adsorption on activated carbon is in most cases an effective method to remove toxic organic matter

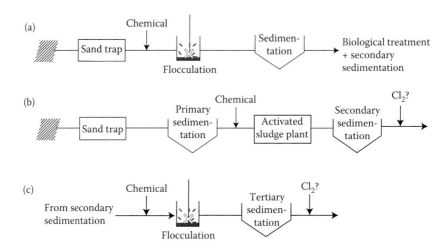

FIGURE 11.1
Precipitation by aluminum sulfate or other aluminum compounds, iron(III) chloride, and calcium hydroxide is able to reduce the phosphorus concentration in wastewater significantly. The precipitation is applied often in combination with mechanical–biological treatment and can be carried out after the sand trap: (a) direct precipitation, before the activated sludge plan; (b) simultaneous precipitation or after the mechanical–biological treatment; and (c) posttreatment.

from industrial wastewater. Finally, extraction can be used to remove both heavy metals and toxic organic compounds, but the method is often relatively expensive and is therefore used less than the other above-mentioned methods. Reverse osmosis can, in principle, be used to remove any pollutant from wastewater. As it is an expensive method, it has not found wide application; however, it is used for the treatment of industrial wastewater, where the amount of wastewater is low but the removal of the toxic substance is important.

11.3 Abatement of Air Pollution by Environmental Technology

The most important air pollution problems are particulate pollution, pollution by carbon monoxide, carbon hydrides, sulfur dioxide, nitrogenous gases, and heavy metals. Furthermore, a number of gases—carbon dioxide (CO_2), methane, and nitrogen oxides—have a greenhouse effect, that is, they will be able to change the climate toward higher atmospheric temperature.

When considering particulate pollution, the source should be categorized with regard to contaminant type. Inert particulates are distinctly different from active solids in the nature and type of their potentially harmful human health effects. Inert particulates comprise solid airborne material, which does not react readily with the environment and does not exhibit any morphological changes as a result of combustion or any other process. Active solid matter is defined as particulate material that can be further oxidized or react chemically with the environment or the receptor. Any solid material in this category can generally, but depending on its composition and size, be considerably more harmful than inert matter of similar size.

A closely related group of emissions are from aerosols of liquid droplets, generally below 5 µm. They can be oil or other liquid pollutants (e.g., Freon) or may be formed by condensation in the atmosphere. Fumes are condensed metals, metal oxides, or metal halides, formed by industrial activities, predominantly as a result of pyrometallurgical processes: melting, casting, or extruding operations. Products of incomplete combustion are often emitted in the form of particulate matter. The most harmful components in this group are particulate polycyclic organic matter (PPOM), which are mainly derivatives of benzo[*a*]pyrene (see Chapter 6).

Natural sources of particulate pollution are sandstorms, forest fires, and volcanic activity. The major sources in towns are vehicles, combustion of fossil fuel for heating and production of electricity, and industrial activity.

Particulate pollution is an important health factor. The most crucial factors are toxicity and size distribution. Many particles are highly toxic, such as asbestos and those of metals such as beryllium, lead, chromium, mercury, nickel, and manganese. In addition, particulate matter is able to absorb gases, which enhances the effects of these components. In this context, the particle size distribution is of particular importance, as particles greater than 10 µm are trapped in the human upper respiratory passage and the specific surface (expressed as m^2 per g of particulate matter) increases with l/d, where d is the particle size. The adsorption capacity of particulate matter, expressed as g adsorbed per g of particulate matter, will generally be proportional to the surface area. Table 11.3 lists some typical particle size ranges.

Particulate pollutants have the ability to adsorb gases, including sulfur dioxide, nitrogen oxides, carbon monoxide, and other toxic gases. The inhalation of these toxic gases is frequently associated with this adsorption, as the gases otherwise would be dissolved in the mouthwash and spittle before entering the lungs.

TABLE 11.3

Typical Particle Size Ranges

	µm
Tobacco smoke	0.01–1
Oil smoke	0.05–1
Ash	1–500
Ammonium chloride smoke	0.1–4
Powdered activated carbon	3–700
Sulfuric acid aerosols	0.5–5

Particulate pollution may be controlled by modifying the distribution pattern—by building taller chimneys, so to speak. This method represents in principle an obsolete philosophy of pollution abatement—dilution, but it is still widely used to reduce the concentration of pollutants at ground level and thereby minimize the effect of air pollution. Particulate control technology can offer a wide range of methods aimed at the removal of particulate matter from gas. These methods are settling chambers, cyclones, filters, electrostatic precipitators, wet scrubbers, and modification of particulate characteristics. The methods with their optimum particle size and efficiency are compared in Table 11.4. The cost of the various installations varies of course from country to country and is dependent on several factors (material applied, standard size or not standard size, automatic, and so on). Generally, electrostatic precipitators are the most expensive solution and are mainly applied for large quantities of air. Wet scrubbers are among the more expensive installations, while settling chamber and centrifuges are the most cost-effective solutions. For more technical details, Jørgensen (2000) could be applied.

ILLUSTRATION 11.1

The concentration of particulate matter at the ground is inversely proportional to the squared height of the chimney.

If the particulate matter has a concentration at the ground of 20 mg/(24 h m^2) and it is required to be less than 0.2 mg/(24 h m^2), how many times taller do we need to build the chimney to meet this requirement?

Solution

The concentration has to be 100 times smaller, which means that the chimney has to be 10 times taller, because 10 times taller implies that the concentration is 10*10 = 100 times smaller.

TABLE 11.4

Characteristics of Particulate Pollution Control Equipment

Device	Optimum Particle Size (μm)	Optimum Concentration (g m^{-3})	Temperature Limitations (°C)	Air Resistance (mm H$_2$O)	Efficiency (% by Weight)
Settling chambers	>50	>100	−30–350	<25	<50
Centrifuges	>10	>30	−30–350	<50–100	<70
Multiple centrifuges	>5	>30	−30–350	<50–100	<90
Filters	>0.3	>3	−30–250	>15–100	>99
Electrostatic precipitators	>0.3	>3	−30–500	<20	<99
Wet scrubbers	>2–10	>3–30	0–350	>5–25	<95–99

A scrubbing liquid, usually water, is used to assist separation of particles, or a liquid aerosol from the gas phase. The operational range for particle removal includes material less than 0.2 μm in diameter up to the largest particles that can be suspended in air. Gases soluble in water are of course also removed by this process.

All types of fossil fuel will produce CO_2 on combustion, which is used in the photosynthetic production of carbohydrates. As such, CO_2 is harmless and has no toxic effect, whatever the concentration levels are, but increased CO_2 concentration in the atmosphere will increase absorption of infrared radiation and the heat balance of the Earth will be changed.

Carbon hydrides are the major components of oil and gas, and incomplete combustion will always involve their emission. By regulation of the ratio of oxygen to fuel, more complete combustion can be obtained, but the emission of carbon monoxide cannot be totally avoided. Motor vehicles are also the major source of carbon monoxide pollution. On average, 1 L of gasoline (petrol) will produce 200 L of carbon monoxide, while it is possible to minimize the production of this pollutant by using diesel instead of gasoline. In most industrial countries, more than 75% of this pollutant, which is very toxic, originates from motor vehicles.

CO_2 pollution is inevitably related to the use of fossil fuels. Therefore, it can only be solved by the use of other alternative (renewable) sources of energy: wind energy, hydropower, solar panels, solar cells, and wave and tide energy. The uses of these forms of energy solve or course the problems associated with the emission of heavy metals and toxic gases by the use of fossil fuel.

Legislation is playing a major role in controlling the emission of carbon hydrides, nitrogen oxides (for details about these gases, see later in this section), and carbon monoxide. As motor vehicles are the major source of these pollutants, control methods should obviously focus on the possibilities of reducing vehicle emission. The methods available today are

1. Motor technical methods
2. Afterburners
3. Alternative energy sources

The first method is based on a motor adjustment according to the relationship between the composition of the exhaust gas and the air/fuel ratio. A higher air/fuel ratio results in a decrease in the carbon hydrides and carbon monoxide concentrations, but to achieve this, a better distribution of the fuel in the cylinder is required, which is only possible through the construction of another gasification system. This method may be considered cleaner technology. At present, two types of afterburners are in use—thermal and catalytic afterburners. In the former type, the combustible material is raised above its autoignition temperature and held there long enough for complete oxidation of carbon hydrides and carbon monoxide to occur. This method is used on an industrial scale (Waid, 1972, 1974) when low-cost purchased or diverted fuel is available; in vehicles, a manifold air injection system is used.

Catalytic oxidation occurs when the contaminant-laden gas stream is passed through a catalyst bed, which initiates and promotes oxidation of the combustible matter at a lower temperature than would be possible in thermal oxidation. The method is used on an industrial scale for the destruction of trace solvents in the chemical coating industry. Vegetable and animal oils can be oxidized at 250–370°C by catalytic oxidation. The exhaust fumes from chemical processes, such as ethylene oxide, methyl methacrylate, propylene, formaldehyde, and carbon monoxide, can easily be catalytically incinerated at even lower temperatures. The application of catalytic afterburners in motor vehicles presents some difficulties due to poisoning of the catalyst with lead. With the decreasing lead concentration in gasoline, it is becoming easier to solve that problem, and the so-called double catalyst system and even three-way catalysts are now finding a wide application. These systems are able to reduce nitrogen oxides and oxidize carbon monoxide and carbon hydrides simultaneously. New catalysts are currently coming into the market and offer a higher efficiency.

Lead in gasoline has been replaced by various organic compounds to increase the octane number. Benzene has been applied but it is toxic and causes air pollution problems because of its high vapor pressure. Methyl tertiary butyl ether (MTBE) is another possible compound for increasing the octane number. It is, however, very soluble and has been found as groundwater contaminant close to gasoline stations. Application of alternative, renewable energy sources is still at a preliminary stage. Most interest has, however, been devoted to electric and hybrid vehicles. It has also been discussed and is under development, to apply hydrogen produced by electricity as fuel. It would produce harmless water by its use as energy source. To eliminate all pollutants, including CO_2, it will of course be necessary to produce the electricity by renewable energy sources.

Fossil fuel contains approximately 2%–5% sulfur (w/w%), which is oxidized by combustion to sulfur dioxide. Although fossil fuel is the major source, several industrial processes produce emissions containing sulfur dioxide, for example, mining, the treatment of sulfur containing ores, and the production of paper from pulp. The total global emission of sulfur dioxide has been decreasing during the last 35 years due to the installation of pollution abatement equipment, particularly in North America, the European Union, and Japan and due to the use of low-sulfur diesel. The concentration of sulfur dioxide in the air is relatively easy to measure and sulfur dioxide has been used as an air pollution indicator. High values recorded by atmospheric inversion are typical.

Sulfur dioxide is oxidized in the atmosphere to sulfur trioxide, which forms sulfuric acid in water. Since sulfuric acid is a strong acid, it is easy to understand that sulfur dioxide pollution indirectly causes corrosion of iron and other metals and is able to acidify aquatic ecosystems (see also Jørgensen, 2000). Clean Air Acts have been introduced in all industrialized countries during the 1970s and 1980s. Table 11.5 illustrates some typical sulfur dioxide emission standards, although these may vary slightly from country to country. Notice how the duration is important (see also the discussion in Chapter 2).

The approaches used to meet the requirements of the acts, as embodied in the standards, can be summarized as follows:

1. Fuel switching from high to low-sulfur fuels (the sulfur concentration varies from source to source and it is furthermore possible by treatment of the fossil fuel to remove the sulfur)
2. Modification of the distribution pattern—use of tall stacks
3. Abandonment of very old power plants, which have a particularly high emission
4. Flue gas cleaning

Desulfurization of liquid and gaseous fuel is a well-known chemical engineering operation. In gaseous and liquid fuels, sulfur either occurs as hydrogen sulfide or can react with hydrogen to form hydrogen sulfide. The hydrogen sulfide is usually removed by absorption in a solution of alcanolamine and then converted to elemental sulfur.

Sulfur occurs in coal both as pyritic sulfur and as organic sulfur. Pyritic sulfur is found in small discrete particles within the coal and can be removed by mechanical means, for example, by gravity separation methods. However, 20%–70% of the sulfur content of coal is present as organic sulfur, which can hardly be removed today on an economical basis. Since sulfur recovery from gaseous and liquid fuels is much easier than from solid fuel, which also has other disadvantages, much research has been and is being devoted to the gasification or liquefaction of coal.

When sulfur is not or cannot be economically removed from fuel oil or coal prior to combustion, removal of sulfur oxides from combustion gases will become necessary for compliance with the stricter air pollution control laws.

TABLE 11.5

SO_2 Emission Standards Applied in EU

Duration	Concentration (ppm)	Comments
Month	0.05	
24 h	0.10	Might be exceeded once a month
30 min	0.25	Might be exceeded 15 times/month

The chemistry of sulfur dioxide recovery presents a variety of choices and the following five methods should be considered:

1. Adsorption of sulfur dioxide on active metal oxides with regeneration to produce sulfur.
2. Catalytic oxidation of sulfur dioxide to produce sulfuric acid.
3. Adsorption of sulfur dioxide on charcoal with regeneration to produce concentrated sulfur dioxide.
4. Reaction of dolomite or limestone with sulfur dioxide by direct injection into the combustion chamber. A lime slurry is injected into the flue gas beyond the boilers.
5. Fluidized bed combustion of granular coal in a bed of finely divided limestone or dolomite maintained in a fluid-like condition by air injection. Calcium sulfite is formed as a result of these processes.
6. Washing of the smoke by a scrubber using calcium hydroxide slurry.

Particularly the three latter methods have found wide application particularly to large industrial installations. It is possible to recover the sulfur dioxide or elemental sulfur from these processes, making it possible to recycle the spent sorbing material.

Seven different compounds of oxygen and nitrogen are known: nitrous oxide (N_2O), nitrogen oxide (NO), nitrogen dioxide (NO_2), nitric oxide (NO_3), nitrogen trioxide (N_2O_3), nitrogen tetroxide (N_2O_4), nitrogen pentoxide (N_2O_5)—often summarized as NO_x. From the point of view of air pollution, mainly NO and NO_2 are of interest. The major sources of the two gases are NO—from combustion of gasoline and oil: NO_2—from combustion of oil, including diesel oil.

The emission from motor vehicles can be reduced by the same methods as mentioned for carbon hydrides and carbon monoxide. The air/fuel ratio determines the concentration of pollutants in the exhaust gas. An increase in the ratio will reduce the emission of carbon hydrides and carbon monoxide, but unfortunately will increase the concentration of nitrogenous gases. Consequently, the selected air/fuel ratio will be a compromise. A double catalytic afterburner is applied today. It is able to reduce nitrogenous gases and simultaneously oxidize carbon hydrides and carbon monoxide. The application of alternative energy sources will, as for carbon hydrides and carbon monoxide, be a very useful control method for nitrogenous gases at a later stage.

Nitrogenous gases in reaction with water form nitrates, which are washed away by rain water. In some cases, this can be a significant source of eutrophication. For a shallow lake, for example, the increase in nitrogen concentration due to the nitrogen input from rain water will be rather significant. In a lake with a depth of 1.7 m and an annual precipitation of 600 mm, which is normal in many temperate regions, the annual input will be as much as 0.3 mg/L.

The rapid growth in industrial production during recent decades has exacerbated the industrial air pollution problem, but, due to increased application of continuous processes, recovery methods, air pollution control, use of closed systems, and other technological developments, industrial air pollution has, in general, not increased in proportion to production.

Industry displays a wide range of air pollution problems related to a large number of chemical compounds in a wide range of concentrations.

Since industrial air pollution covers a wide range of problems, it is not surprising that all the three classes of pollution control methods mentioned previously have found application: modification of the distribution pattern, alternative (cleaner) production methods, and particulate and gas/vapor control technology.

All the methods mentioned earlier are also applied for industrial air pollution control. In gas and vapor technology, a distinction has to be made between condensable and noncondensable gaseous pollutants. The latter must usually be destroyed by incineration, while the condensable gases can be removed from industrial effluents by

- Absorption
- Adsorption
- Condensation
- Combustion

Recovery is feasible by the first three methods.

Absorption is a diffusion process that involves the mass transfer of molecules from the gas state to the liquid state along a concentration gradient between the two phases. Absorption is a unit operation that is enhanced by all the factors generally affecting mass transfer; that is, high interfacial area, high solubility, high diffusion coefficient, low liquid viscosity, increased residence time, turbulent contact between the two phases, and possibilities for reaction of the gas in the liquid phase. This last factor is often very significant and almost 100% removal of the contaminant is the result of such a reaction. Acidic components can be easily removed from gaseous effluents by absorption in alkaline solutions, for instance, sulfur dioxide and nitrogen oxides. Correspondingly, alkaline gases can easily be removed from the effluent by absorption in acidic solutions. Table 11.6 gives an overview of applicable absorber reagents.

CO_2, phenol, and hydrogen sulfide are readily absorbed in alkaline solutions in accordance with the following processes:

$$CO_2 + 2NaOH \rightarrow 2Na^+ + CO_3^{2-} \qquad (11.1)$$

$$H_2S + 2NaOH \rightarrow 2Na^+ + S^{2-} + 2H_2O \qquad (11.2)$$

$$C_6H_5OH + NaOH \rightarrow C_6H_5O^- + Na^+ + H_2O \qquad (11.3)$$

TABLE 11.6

Absorber Reagents

Reagents	Applications
$KMnO_4$	Rendering, polycyclic organic matter
NaOCl	Protein adhesives
Cl_2	Phenolics, rendering
Na_2SO_3	Aldehydes
NaOH	CO_2, H_2S, phenol, Cl_2, pesticides
$Ca(OH)_2$	Paper sizing and finishing
H_2SO_4	NH_3, nitrogen bases

Ammonia is readily absorbed in acidic solutions:

$$2NH_3 + H_2SO_4 \rightarrow 2NH_4^+ SO_4^{2-} \tag{11.4}$$

Adsorption is the capture and retention of a component (adsorbate) from the gas phase by the total surface of the adsorbing solid (adsorbent). In principle, the process is the same as with wastewater treatment; the theory is equally valid for gas adsorption. Adsorption is used to concentrate (often 20–100 times) or store contaminants until they can be recovered or destroyed in the most economical way. These processes are often described as either Langmuir's or Freundlich's adsorption isotherms, which are also applied for a quantitative description of this process in nature. Adsorption is dependent on temperature: increased temperature means that the molecules move faster and therefore it is more difficult to adsorb them. There are four major types of gas adsorbents, the most important of which is activated carbon, but also aluminum oxide (activated aluminum), silica gel, and zeolites are used.

The selection of adsorbent is made according to the following criteria:

1. High selectivity for the component of interest.
2. Easy and economical to regenerate.
3. Availability of the necessary quantity at a reasonable price.
4. High capacity for the particular application, so that the unit size will be economical. Factors affecting capacity include total surface area involved, molecular weight, polarity activity, size, shape, and concentration.
5. Pressure drop, which is dependent on the superficial velocity.
6. Mechanical stability in the resistance of the adsorbent particles to attrition. Any wear and abrasion during use or regeneration will lead to an increase in bed pressure drop.

7. Microstructure of the adsorbent should, if at all possible, be matched to the pollutant that has to be collected.

8. The temperature, which has a profound influence on the adsorption process, as already mentioned. Regeneration of the adsorbents is an important part of the total process. A few procedures are available for regeneration.

Although the regeneration is 100%, the capacity of the adsorbent may be reduced 10%–25% after several regeneration cycles due to the presence of fine particulates and/or high-molecular-weight substances, which cannot be removed in the regeneration step.

Combustion is defined as rapid, high-temperature gas-phase oxidation. The goal is the complete oxidation of the contaminants to CO_2 and water (H_2O), sulfur dioxide, and nitrogen dioxide. The process is often applied to control odors in rendering plants, paint and varnish factories, rubber tire curing, and petrochemical factories. It is also used to reduce or prevent an explosion hazard by burning any highly flammable gases for which no ultimate use is feasible. The efficiency of the process is highly dependent on temperature and reaction time, and also on turbulence or the mechanically induced mixing of oxygen and combustible material. The relationship between the reaction rate, r and the temperature can be expressed by Arrhenius' equation (see Chapter 9).

Heavy metals, which may be defined as the metals with a specific gravity >5.00 kg/L, comprise 70 elements. Most of them are, however, only rarely found as pollutants. The heavy metals of environmental interest form almost insoluble compounds with sulfide and phosphate and form very stable complexes with many ligands present in the environment. It means, fortunately, that most of the heavy metals are not very bioavailable in most environments (see also the presentation of bioremediation in Chapter 12 and the overview of heavy metals presented in Chapter 6).

A number of enzymes activated by metal ions and metalloenzymes are known. The first mentioned group comprises iron, cobalt, chromium, vanadium, and selenium. Copper, zinc, iron, cobalt, and molybdenum are able with a stronger bond to form metalloenzymes: metalloproteins, metalloporphyrines, and metalloflavines. As pollutants are particularly lead, cadmium, and mercury in focus, because of their extremely high toxicity (see the details in Chapter 6).

Heavy metals are emitted to the atmosphere by energy production and a number of technological processes (see Table 11.7). It makes the atmospheric deposition of heavy metals, originated from human activities, the dominant pollution source for the vegetation of natural ecosystems—forests, wetlands, peatlands, and so on. The heavy metal content in sludge and fertilizers plays a more important role for agricultural land where the inputs of heavy metals by irrigation, natural fertilizers, and application of chemicals, including pesticides, may add to the overall pollution level. The atmosphere and

TABLE 11.7

Important Atmospheric Pollution Sources of Heavy Metals[a]

Source	Heavy Metals
Incineration of oil	V, Ni
Incineration of coal	Hg, V, Cr, Zn, As
Gasoline	Pb (leaded gasoline)
Metal industry	Fe, Cu, Mn, Zn, Cr, Pb, Ni, Cd, and others
Application of pesticides	Hg, Cr, Cu, As
Incineration of solid waste	Hg, Zn, Cd, and others

[a] See also Chapter 3.

hydrosphere have both a well-developed ability for "self-purification"—for heavy metals by removal processes, for instance, sedimentation. The lithosphere has a high buffer capacity toward the effects of most pollutants, and has an ability of self-purification, for instance, by runoff and uptake by plants, although the rates usually are much lower than in the two other spheres.

11.4 Solution of Solid Waste Problems by Environmental Technology

Solid wastes include an incredible miscellany of items and materials, which makes it impossible to indicate one simple solution to the problem. It is necessary to apply a wide spectrum of solutions to the problems according to the source and nature of the waste. Table 11.8 shows a classification of

TABLE 11.8

Classification of Solid Waste

Type of Waste	kg/Inhabitant/day (Approximate)
Domestic garbage	4.6
Agricultural waste[a]	12
Mining waste[b]	17.5
Wastes from construction[c]	0.13
Industrial waste	1.4
Junked automobiles	0.15

[a] Not mentioned further in this context. Recycling is recommended and also widely in use.

[b] A substantial part is used for land filling.

[c] Not mentioned further in this context. Recycling is possible and recommended, but part is also used for land filling.

solid waste. Typical quantities per inhabitant in a technological society are included. The numbers represent the situation in EU 2010 and are taken from various sources.

A universal method for the treatment of solid waste does not exist and it is necessary to analyze each individual case to find a relevant solution to the problem. The analysis begins with an examination of mass flows to ascertain whether there is an economical basis for reuse (e.g., bottles), recovery of valuable raw materials (paper and metals), or utilization of organic matter, as a soil conditioner or for the production of energy. These considerations are illustrated in Figure 11.2. Mass balances for important materials must therefore be set up, including toxic waste.

The analysis of mass flows is the framework for a feasible solution, which takes economic as well as environmental issues into consideration. However, legislation and economical means are required to achieve the most obvious management goals. For instance, the use of returnable bottles can be realized either by banning the use of throw-away bottles or by placing a purchase tax on throw-away bottles and not on returnable bottles. The management problems are highly dependent on the technological methods used for solid waste treatment. Methods based on cleaner technology (see also Chapter 13) and recycling, to reduce the amount of waste produced, often require environmental legislation or economical means (green tax) to guarantee success. Methods based on deposition will require a comprehensive knowledge of environmental effects on the ecosystems involved. Complete decomposition to harmless components is not possible for most solid waste. For instance, incineration will produce a slag, which might cause a deposition problem, and smoke, which involves air pollution.

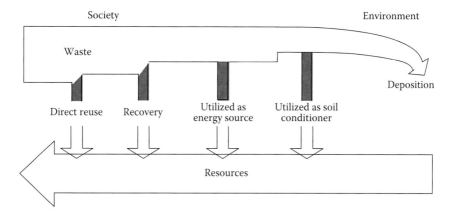

FIGURE 11.2
Principles of recycling illustrated by mass flows.

11.4.1 Methods for Treatment of Solid Waste: An Overview

Table 11.9 provides an overview of the methods applicable to solid waste. The table indicates on which principle the method is based—recycling, deposition, or decomposition to harmless components. It also shows the type of solid waste the method is able to treat.

The solutions of the solid waste problems can often be classified according to the source: (1) sludge, (2) domestic garbage, (3) industrial, mining, and hospital waste, and (4) agricultural waste. The methods used for the four classes are, to a certain extent, different, although there is some overlap, among them. The highest content of toxic substances is found in type 3 waste, and Table 11.9 indicates clearly which methods can be applied for type 3 waste. The method often has to be tailored to the toxic compound content of the waste. Later in this section, applicable solutions for toxic waste are presented and these methods should be used in case the solid waste—whatever the source is—contains harmful amounts of toxic substances. Generally, the three Rs should be the preferred, most environmentally friendly method: reduce, reuse, and recycle. A reduction of the toxic effect should be the key issue and the methods presented last in this section should be seriously considered. The presence of toxic substances in sufficiently high concentrations may even prevent the use of composting, land filling, pyrolysis, aerobic, and anaerobic treatment methods, because the microbiological processes based on these method are inhibited by toxic substances.

TABLE 11.9

Methods for Treatment of Solid Waste

Method	Principle	Applicable to
Conditioning and composting	Decomposition before deposition	Sludge, domestic garbage, agricultural waste
Anaerobic treatment	Decomposition, deposition, utilization of biogas	Sludge, agricultural waste including manure
Thickening, filtration, centrifugation, and drying	Dewatering before further treatment and deposition	Sludge that may contain toxic substances
Combustion and incineration	Decomposition and utilization of energy	All types of waste
Separation	Recycling, partially	Domestic garbage
Dumping ground	Decomposition, deposition	Domestic garbage
Pyrolysis	Recycling, utilization of energy, decomposition	Domestic garbage
Precipitation and filtration	Deposition or recycling	Industrial waste
Land filling	Deposition	Domestic garbage and agricultural waste
Aerobic treatment	Decomposition and deposition	Sludge, agricultural waste
Recovery	Precipitation, absorption, adsorption, and distillation	Industrial and mining waste with high toxicity

11.4.1.1 Sludge

In most wastewater treatment, the impurities are not actually removed, but rather concentrated in the form of sludge. Only when a chemical reaction takes place does the real removal of the impurities occur, for example, by chemical or biochemical oxidation of organics to CO_2 and H_2O, or denitrification of nitrate to nitrogen gas. Sludge from industrial wastewater treatment units in most cases requires further concentration before its ultimate disposal. In many cases, two- or even three-step processes are used to concentrate the sludge. It is often an advantage to use further thickening by gravity, followed by such treatments as filtration or centrifugation. The characteristics of a sludge are among the factors that influence the selection of the best sludge treatment method. The sludge characteristics vary with the wastewater and the wastewater treatment methods used. One of the important factors is the concentration of the sludge. The specific gravity of the sludge is another important factor, since the effect of gravity is utilized in the thickening process. The specific gravity of activated sludge increases linearly with the sludge concentration. This corresponds to a specific gravity of 1.08 g/mL for actual solids. The water can be removed from a sludge by such processes as vacuum filtration, centrifugation, and sand-bed drying.

When sludge is being considered for use as a soil conditioner, its chemical properties are of prime importance. The nutrient content (nitrogen, phosphorus, and potassium), in particular, is of interest. Furthermore, knowledge of the heavy metals and organic contaminants in sludge is very important because of their toxicity. Standards for sludge applicable as soil conditioner are used in most industrialized countries (see Jørgensen, 2000). Finally, it must be considered that normal wastewater treatment processes, such as sedimentation, chemical precipitation, and biological treatments, remove considerable amounts of pathogens that are concentrated in the sludge. A significant reduction in the number of pathogenic organisms has been found to occur during anaerobic digestion, but they are not destroyed entirely. Combustion, intensive heat treatment of sludge, or treatment with calcium hydroxide would of course eliminate the hazard of pathogenic microorganisms.

Sludge conditioning is a process that alters the properties of the sludge to allow the water to be removed more easily. The aim is to transform the amorphous gel-like sludge into a porous material, which will release water. Conditioning of the sludge can be accomplished by either chemical or physical means. Chemical treatment usually involves the addition of coagulants or flocculants to the sludge. Typical doses of inorganic coagulants, such as aluminum sulfate, iron(III) chloride, and calcium hydroxide, are as much as 20% of the weight of the solid, while a typical dose of organic polymer is less than 1% of the weight of the solid. This does not necessarily mean that the cost of using synthetic polymers is lower, since the polymers cost considerably more per kg than the inorganic chemicals used as conditioners. The amorphous gel-like structure of the sludge is destroyed by heating. The filtration rate of activated sludge is increased more than a 1000-fold after heat treatment. Typically, the heat-treatment conditions are 30-min treatments at 150–200°C under a pressure of 10–15 atmospheres. A great advantage of heat conditioning is, of course, that the pathogens are destroyed. Conditioning by freezing has also been reported. Thickening of sludge is possible by filtration. The filter area is determined from the specific sludge resistance and the amount to be treated by filtration. As the specific resistance of conditioned sludge may vary considerably, it is recommended to design for a surplus capacity of at least 25%–50%.

Several types of filters are applicable to achieve a high concentration of dry matter (30%–45%) in the sludge, which will reduce the cost of transportation significantly. Vacuum filtration is used to remove water from sludge by applying a vacuum across a porous medium. As the rotary drum passes through the slurry in the slurry tank, a cake of solids is built on the drum surface and the water is removed by vacuum filtration through the porous medium on the drum surface. The cake is removed from the drum by a knife edge. Often the porous filter is washed with water before it is returned to the slurry tank.

Filter presses (mainly plate and frame type) are more cost effective than vacuum filters and are more widely used. Calcium hydroxide slurry is often used as a filter medium. Filter presses operate by charge basis, as the filter cakes have to be removed when the filtration rate is reduced to a certain level.

Filter belts, are also widely used. They have the advantage that they operate continuously, but usually they provide a lower concentration of dry matter than the filter press (about 30%–35% versus 40% for filter presses).

The filtrate from the treatment of municipal sludge has a high ammonium-nitrogen concentration, which may be advantageous to remove by stripping to eliminate the nitrogen loading in the wastewater, as the filtrate must always be recycled.

Centrifugation can also be used in the removal of water from wastewater sludge. Of the various types of centrifuge, the solid-bowl centrifuge is considered to offer the best clarification and water-removal properties. The use of electrolytes will increase the recovery at a given flow rate or increase the flow rate for a given recovery.

If the sludge contains biodegradable organic material, it may be advantageous to treat the sludge by aerobic or anaerobic digestion before thickening. Anaerobic digestion is by far the most common method of treating municipal sludge. It creates good conditions for the growth of microorganisms. The end products of anaerobic digestion are CO_2 and methane. The temperature is commonly set at about 35°C in order to maintain optimum conditions in the digester. Unfortunately, anaerobic digestion results in considerable quantities of nutrients going into solution (Dalton et al., 1968), which means that a significant amount of nutrient material will be returned to the treatment plant if the supernatant is separated from the sludge. The principal function of anaerobic digestion is to convert as much of the sludge as possible to end products: liquids and gases. Anaerobic decomposition generally produces less biomass than aerobic processes. Heavy metal ions may slow down the decomposition rate. The presence of ammonia and hydrogen sulfide may have the same effect (Vavilin et al., 1995).

The microorganisms can be divided into two broad groups: *acid formers* and *methane formers*. The *acid formers* consist of facultative and anaerobic bacteria, and soluble products are formed through hydrolysis. The soluble products are then fermented to acids and alcohols of lower molecular weight. The *methane formers* are strictly anaerobic bacteria that convert the acids and alcohols, along with hydrogen and CO_2, to methane. The biogas produced by anaerobic digestion of sludge from wastewater treatment is often utilized at wastewater treatment plants for the production of heat and for the production of electricity (a gas turbine is used). The amount of gas is usually not sufficient to make the sale of gas possible.

The properties of aerobically digested sludge are similar to those of anaerobically digested sludge. An advantage is that some of the operational problems attending anaerobic digesters are avoided, but the disadvantage compared with anaerobic digestion is that the process is more expensive since oxygen must be provided and energy recovery from methane is not possible. Since aerobic digestion is less used in industrial wastewater processes than in the treatment of municipal wastewater, it is not appropriate here to go into further detail. The purpose of drying sludge is to prepare it for use as a soil conditioner or for incineration. Air drying of the sludge on sand beds is often used to reach a moisture content of about 90%. Also, such drying techniques as flash drying and rotary drying are used to remove water from sludge. Often waste heat from the incineration process itself is used in drying. However, the cost of combined drying and combustion is higher than the cost of incineration alone.

11.4.1.2 Domestic Garbage

The composition of domestic garbage varies from country to country. To a certain extent, the environmental legislation and the economic level of the nation is reflected in this composition. The amount of domestic garbage per inhabitant is increasing. In most developed countries, the growth was 2%–4% from the mid-1950s to 1963, while it has been lower in the period from the mid-1970s to 2010 (1%–2%).

It is often advantageous to carry out a comprehensive analysis of the domestic garbage in order to estimate the economic and ecological consequences of various environmental management strategies, including the selection of the most relevant treatment methods. The analysis in this case will include the following items: (1) food waste, (2) paper, (3) textiles, (4) leather and rubber, (5) plastics, (6) wood, (7) iron, (8) aluminum, (9) other metals, (10) glass and ceramic products, (11) ash and dust, (12) stone, (13) garden waste, and (14) other types of waste.

Separation of solid waste can be achieved either in a central plant or by the organization of separate collection of paper, glass, metals, and other types of domestic waste. In many industrialized countries, the separate collection of toxic compounds is compulsory—not only from industries but also from households.

Dumping grounds (landfills) were previously the most common handling method for solid waste. Today, it is mainly in use in smaller towns, often after grinding or compression, which reduces the volume by 60%–80%.

Deposition of solid waste on a dumping ground is an inexpensive method, but it has a number of disadvantages:

1. Possible contamination of groundwater
2. Causes inconveniences due to the smell
3. Attracts noxious animals, such as flies and rats

During the deposition, several processes take place:

1. Decomposition of biodegradable material
2. Chemical oxidation of inorganic compounds
3. Dissolution and washout of material
4. Diffusion processes

Where the decomposition takes place in the aerobic layers, CO_2, water, nitrates, and sulfates are the major products liberated, while decomposition in anaerobic layers leads to the formation of CO_2, methane, ammonia, hydrogen sulfide, and organic acids.

Water percolating from dumping grounds has a very high concentration of BOD and nutrients and frequently also of toxic compounds. Therefore, it cannot be discharged into receiving waters. If it cannot be recycled on the dumping ground, this wastewater must be subject to some form of treatment.

Composting has been applied as a treatment method in agriculture for thousands of years. The method is still widely applied for the treatment of agricultural waste. Organic matter from untreated solid waste cannot be utilized by plants, but it is necessary to let it undergo a degree of biological decomposition through the action of microorganisms.

Again, we can distinguish between aerobic and anaerobic processes. A number of factors control these processes:

1. The ratio of aerobic to anaerobic processes, which, of course, is determined by the available oxygen (diffusion process).
2. Temperature. Different classes of microorganisms are active within different temperature ranges (see Table 11.10). Heat is produced by the decomposition processes. The thickness of the layer determines

TABLE 11.10

Classification of Microorganisms

	Temperature Optimum	Temperature Range
Psychrophile	15–20°C	0–30°C
Mesophile	25–35°C	10–40°C
Thermophile	50–55°C	25–80°C

to what extent this heat can be utilized to maintain a temperature of 60–65°C, which is considered to be the optimum. In this context, it is also important to mention that composting at a relatively high temperature means a substantial reduction in the number of pathogenic microorganisms and parasites.

3. The water content should be 40%–60%, as this gives the optimum conditions for the processes of decomposition.

4. The carbon-to-nitrogen (C/N) ratio should be in accordance with the optimum required by the microorganisms. Domestic garbage has a C/N ratio of 80 or more due to the high content of paper, while the optimum for composting is 30. It is therefore advantageous to add sludge from the municipal sewage plant to adjust the ratio to about 30. Sludge usually has a C/N ratio of 10 or even less. During composting, the C/N ratio is decreased as a result of respiration, which converts some the organic matter to CO_2 and H_2O.

5. The optimum conditions for the microorganisms include a pH around 6 (6–8). Usually, pH increases as a result of the decomposition processes. If the pH is too low, calcium hydroxide should be added and if the pH is too high, sulfur should be added. Sulfur activates sulfur bacteria, which produce sulfuric acid.

A composting plant generally applies composting in containers followed by composting in stacks. Composting in containers with the addition of *air accelerates* the decomposition processes, but this process might be excluded in smaller plants. The compost is used as soil conditioner when it is stable, that is, the decomposition processes are almost terminated, which may be determined by a respirometric technique. The standards for sludge are also applied for compost.

Incineration of domestic garbage and solid waste, in general, is very attractive from a sanitary point of view, but it is a very expensive method, which has some environmental disadvantages. Valuable material such as paper is not recycled, a slag, which must be deposited, is produced and air pollution problems are involved. Nevertheless, incineration has been the preferred method in many cities and larger towns. This development might be explained by the increasing amount of combustible material in domestic garbage and the possibilities of combining incineration plants with district heating. The optimum combustion temperature is 800–1050°C. If the temperature is below this range, incomplete combustion will result (dioxines may be produced) and at a higher temperature the slag will melt and prevent even air distribution. The composition of the solid waste determines whether this combustion temperature can be achieved without using additional fossil fuel. If the ash content is less than 60%, the water content is less than 50%, and the combustible material more than 25%, no additional fossil fuel is required.

The composition of the slag and the fly ash, which is collected in the filter, is, of course, dependent on the composition of the solid waste. It is notable that a minor amount of the slag and the fly ash always will be unburned material. Slag and fly ash comprise 25%–40% of the weight of the solid waste. The slag will have a high concentration of heavy metals in most cases but they are usually chemically bound as very water insoluble oxides. Although all modern incineration plants have filters, the air pollution problem is not completely solved. Hydrogen chloride, in particular, can cause difficulties, because it is toxic and highly corrosive. It results from incineration of chlorine compounds among which polyvinyl chloride (PVC) is the most important.

Pyrolysis is the decomposition of organic matter at elevated temperature without the presence of oxygen. For pyrolysis of solid waste, a temperature of 850–1000°C is generally used. The process produces gas and a slag, from which metals can be easily separated due to the anoxic atmosphere. Pyrolysis is a relatively expensive process, although slightly less expensive than incineration. The smoke problems are the same as those of incineration, but the easy recovery of metals from the slag is, of course, an advantage. *The gas produced has a composition close to coal gas* and in most cases can be used directly in the gas distribution system without further treatment. Approximately 500 m^3 of gas are produced per ton of solid waste, but 300–400 m^3 are used in the pyrolysis process to maintain the temperature.

Comparison of the methods applied for domestic garbage is presented in Table 11.11. Composting is the process that is most acceptable from an ecological point of view for treatment of solid waste, including domestic garbage, provided of course that toxic substances in harmful concentrations can be excluded. This is according to the six waste treatment principles that will be mentioned in Chapter 14 where the integrated ecological–environmental management will be discussed. Furthermore, in accordance with Chapter 12, that focuses on ecological engineering methods, composting may be designated as ecological engineering method. Pyrolysis also gives the opportunity to reuse waste products as raw materials or energy source. When the waste heat from incineration plants is utilized, this method of course becomes more attractive when evaluated according to the ecological principles.

Composting is, however, less area-intensive than incineration and pyrolysis, although it will require relatively less area than landfills, especially when compressing and grinding are used to reduce the volume of domestic and industrial garbage. Incineration is the most widely used method for

TABLE 11.11

Comparison of Methods for Treatment of Solid Waste, Particularly Domestic Garbage without Harmful Concentrations of Toxic Substances

Method	Resource Friendly	Ecological Evaluation	Area Needed	Cost per kg Waste
1. Dumping ground	Unacceptable	Unacceptable	Large	Low
2. Dumping ground with pretreatment	About 50% unacceptable	Slightly but still unacceptable	Better than 1	More than 1
3. Composting	Acceptable	Acceptable	Medium	70%–85% of 4 and 5
4. Pyrolysis almost	Almost acceptable	Acceptable	Low	High
5. Incineration	Unacceptable, unless heat is recovered	Unacceptable	Low	High

treatment of solid waste in bigger towns and cities. The relatively small area required per kg of waste is the dominant factor here.

The cost of the treatment of course varies with the size of the plant and the composition of waste. Generally, the treatment costs increases in the following order of treatment methods: dumping ground → dumping ground with compressing and grinding of waste → composting → incineration or pyrolysis.

11.4.1.3 Industrial, Mining, and Hospital Waste

This type of waste causes particular problems because it may contain toxic matter in relatively high concentrations. Hospital wastes are especially suspect because of contamination by pathogens and the special waste products, such as disposable needles and syringes and radioisotopes used for detection and therapy. The waste from industry and mining varies considerably from place to place, and it is not possible to provide a general picture of its composition. If the composition permits, it can be used for land filling, but if it contains toxic matter special treatment will inevitably be required. The composition may be close to that of domestic garbage in which case the treatment methods mentioned above can be applied.

Hospital solid waste is being studied in only a few locations to provide data on current practices and their implications. Most hospital waste is now incinerated and this might, in many cases, be an acceptable solution. However, if the solid waste contains toxic matter, it should be treated along the lines given for industrial wastes below. At least, waste from hospital laboratories should be treated as other types of chemical waste. It means that medicine residues should be collected and be treated as toxic waste. Unused medicine can, in many European countries, be handed over to the drugstores for further treatment. Furthermore, there has been an increased interest in the treatment of wastewater from hospitals, and it will inevitably produce sludge that has to be treated as chemical, toxic waste, and may of course contain medicine.

Industrial and mining waste containing toxic substances, such as heavy metals or toxic organic compounds, should be treated separately from other types of solid waste, which means that it cannot be treated by the methods mentioned above. A number of countries have built special plants to handle this type of waste, which could be called chemical waste. Such plants may include the following treatment methods:

1. Combustion of toxic organic compounds. The heat produced by this process might be used for district heating. Organic solvents, which are not toxic, should be collected for combustion, because discharge to the sewer might overload the municipal treatment plant. For example, acetone is not toxic to biological treatment plants, but 1 kg of acetone uses 2.2 kg of oxygen in accordance with the following process:

$$(CH_3)_2CO + 4O_2 \rightarrow 3CO_2 + 3H_2O \tag{11.5}$$

   A different combustion system might be used for pumpable and nonpumpable waste.

2. Compounds containing halogens or sulfur should be treated only in a system that washes the smoke to remove the hydrogen halogenide or sulfuric acid formed.

3. Waste oil can often be purified and the oil reused.

4. Waste containing heavy metals requires deposition under safe conditions after a suitable pretreatment. Recovery of precious metals is often economically viable and is essential for mercury because of its high toxicity. The pretreatment consists of a conversion to the most relevant oxidation state for deposition, for example, chromates should be reduced to chromium in oxidation state 3. Furthermore, metals should be precipitated as the very insoluble hydroxides (see also Section 11.1) before deposition.

5. Recovery of solvents by distillation becomes increasingly attractive from an economical point of view due to the growing costs of oil products, as most solvents are produced from mineral oil.

It is finally recommended to examine the possibilities of recovering plastics. The main problem with recycling plastics is associated with the wide range of chemically different types of plastic. A mixture of different plastic types that may be recovered does not have the same properties as the particular plastic types, and can therefore only be used where the strength and the color is of minor importance, for instance, for plastic bags.

11.4.1.4 Agricultural Waste

Recycling is used to a great extent for many types of agricultural waste. For instance, manure and dunghills are used as a natural fertilizer. The problem is, however, that although manure contains nutrients useful for application as natural fertilizer, if it is used without sufficient control and too intensively, the nutrients will inevitably pollute rivers and lakes and even coastal areas, due to the organic matter and nutrients of the manure. Therefore, there are limits in many countries on how much manure is permitted per hectare. If it is a problem, however, the use of ecotechnology should be considered as it is suited to nonpoint sources of pollution. Recently, manure has been increasingly used for production of biogas, which is practically the same method as the previously described anaerobic digestion of sludge. Manure may contain veterinary medicine, including antibiotics, which is, of course, unacceptable to spread in nature and is inevitable when manure is used as a natural fertilizer (see Jørgensen 2011).

Straw left on the field after harvest of grains was previously incinerated on the field, but this practice is no longer allowed in many countries due to the danger of fire. Straw is, however, increasingly used as fuel for heating. A possibility, of course, is also to plough the straw in the soil and thereby be able to utilize the carbon and nutrients of the straw. Agricultural waste will be discussed further in Chapter 12 dealing with ecotechnology.

References

Dalton, F., E. Stein, and H.T. Lyman. 1968. Land reclamation—A complete solution to the sludge and solids disposal problem. *Journal of the Water Pollution Control Federation* 40: 789–801.

Hahn, H.H. and N. Muller. 1995. Factors affecting water quality of (large) rivers—Past-experiences and future outlook. In V. Novotny and L. Somlyódy (eds.), *Remediation and Management of Degraded River Basins with Emphasis on Central and Eastern Europe*. Springer-Verlag, Berlin, Germany, 385–426.

Henze, M. and H. Ødegaard. 1995. Wastewater treatment process development in Central and Eastern Europe—Strategies for a stepwise development involving chemical and biological treatment. In: V. Novotny and L. Somlyódy (eds.), *Remediation and Management of Degraded River Basins with Emphasis on Central and Eastern Europe*. Springer-Verlag, Berlin, Germany, 357–384.

Jørgensen, S.E. 2000. *Principles of Pollution Abatement*. Elsevier, Amsterdam, 520pp.

Jørgensen, S.E. (ed.) 2011. *Handbook of Ecological Models Used in Ecosystem and Environmental Management*. CRC Press, Boca Raton, 620pp.

Jørgensen, S.E., J.C. Marques, and S.N. Nielsen. 2015. *Integrated Environmental Management: A Transdisciplinary Approach*. CRC Press, Boca Raton, 380pp.

Loosdrecht, M. van. 1998. Upgrading of waste water treatment process for integrated nutrient removal—The BCFS© process. *Water Science and Technology* 37: 234–256.

Novotny, V. and L. Somlyódy (eds.). 1995. *Remediation and Management of Degraded River Basins with Emphasis on Central and Eastern Europe.* Springer-Verlag, Berlin, Germany.

Vavilin, V.A. et al. 1995. Modelling ammonia and hydrogen sulfide inhibition in anaerobic digestion. *Water Research* 29(3): 827–835.

Waid, D.E. 1972. Controlling pollutants via thermal incineration. *Chemical Engineering Progress* 68: 57–58.

Waid, D.E. 1974. *Thermal Oxidation or Incineration.* Pollution Control Association, Pittsburgh, Pennsylvania, pp. 62–79.

12

Application of Ecotechnology/Ecological Engineering in Environmental Management

12.1 What Is Ecotechnology and Ecological Engineering?

H.T. Odum was among the first to define ecological engineering (Odum, 1962, 1983) as the "environmental manipulation by man using small amounts of supplementary energy to control systems in which the main energy drives are coming from natural sources." Odum further developed the concept (Odum, 1983) of ecological engineering as follows: ecological engineering, the engineering of new ecosystems designs, is a field that uses systems that are mainly self-organizing. Straskraba (1985) has defined ecological engineering, or as he calls it ecotechnology, more broadly, as the use of technological means for ecosystem management, based on a deep ecological understanding, to minimize the costs of measures and their harm to the environment. Ecological engineering and ecotechnology may be considered synonymous.

Mitsch and Jørgensen (1989) give a slightly different definition, which, however, covers the same basic concept as the definition given by Straskraba and also encompasses the definition given by H.T. Odum. They define ecological engineering and ecotechnology as the design of human society with its natural environment for the benefit of both. It is engineering in the sense that it involves the design of man-made or natural ecosystems or parts of ecosystems. Like all engineering disciplines, it is based on basic science, in this case ecology and system ecology. The biological species are the components applied in ecological engineering. Ecological engineering represents therefore a clear application of ecosystem theory.

Ecotechnic is another often-applied word, but also it encompasses in addition to ecotechnology or ecological engineering the development of technology applied in society, based upon ecological principles, for instance, all types of cleaner technology, particularly if they are applied to solve an environmental problem. Even the use of composting, solar cells, and wind energy may be considered in some context as ecotechnique.

Ecological engineering should not be confused with bioengineering or biotechnology. Biotechnology involves the manipulation of the genetic structure of the cells to produce new organisms capable of performing certain functions. Ecotechnology does not manipulate at the genetic level, but operates at several steps higher in the ecological hierarchy. The manipulation takes place on an assemblage of species and/or their abiotic environment as a self-designing system that can adapt to changes brought about by outside forces, controlled by human or by natural forcing functions.

Ecological engineering is *not* the same as environmental engineering, which is involved in cleaning up processes to prevent pollution problems— it is engineering applying environmental technology, which is covered in Chapter 11. It uses settling tanks, filters, scrubbers, and man-made components, which have nothing to do with the biological and ecological components that are applied in ecological engineering, although the use of environmental engineering aims toward reducing man-made forcing functions on ecosystems. Environmental engineering uses technological solutions to solve environmental problems, while ecological engineering uses ecological solutions to solve environmental problems. Ecotechnic mentioned above may be considered to include, in addition to ecological engineering, environmental technology based on ecological principles such as technology using recirculation and reuse. Composting of solid waste is ecotechnic, but it is not ecological engineering, because it uses technology to be able to reuse solid waste as compost. The toolboxes of the two types of engineering are completely different. Ecological engineering uses ecosystems, communities, organisms, and their immediate abiotic environment. Environmental engineering uses man-made technological solutions.

All applications of technologies are based on quantification, meaning that a diagnosis is developed before the selection of the solution to the environmental problem is made. Ecosystems are very complex systems and the quantification of their reactions to impacts or manipulations therefore becomes complex. Fortunately, ecological modeling and the use of ecological indicators represent well-developed tools to survey ecosystems, their reactions, and the linkage of their components. Ecological modeling is able to synthesize our knowledge about an ecosystem and makes it possible to a certain extent to quantify any changes in ecosystems resulting from the use of both environmental engineering and ecological engineering (see Chapter 8). Ecological engineering may also be used directly to design constructed ecosystems. Consequently, ecological modeling and ecological engineering are two closely cooperating fields. Research in ecological engineering was originally covered by the *Journal of Ecological Modelling*, which was initially named *Ecological Modelling— International Journal on Ecological Modelling and Engineering and Systems Ecology* to emphasize the close relationship between the three fields. *Ecological Engineering* was launched as an independent journal in 1992, and the name of *Ecological Modelling* was changed to *Ecological Modelling—An International Journal on Ecological Modelling and Systems Ecology*. Meanwhile, the journal *Ecological Engineering* has been successful in covering the field of ecological engineering, which has grown rapidly during the 1990s owing to increasing acknowledgment of the need for other technologies than environmental technology in our efforts to solve pollution problems in an environment-friendly and cost-moderate manner. This development does not imply that ecological modeling and ecological engineering are moving in different directions. On the contrary, ecological engineering is increasingly using models to perform designs of constructed ecosystems or to quantify the results of application of specific ecological engineering methods for comparison with alternative, applicable methods (Luca Palmeri et al., 2013 and Straskraba, 1993).

The next section is devoted to a classification of ecotechnology that will be able to reveal in more detail the applicability of ecotechnology. The following sections of the chapter cover illustrative examples on how to apply ecotechnology for treatment of agricultural waste, for treatment of wastewater by the use of waste stabilization ponds and constructed wetlands, and for soil remediation with particular emphasis on examples of reducing toxic substance concentrations.

12.2 Classification of Ecological Engineering/Ecotechnology

Ecotechnology may be based on one or more of the following four types or classes:

1. Ecosystems are used to reduce or solve a pollution problem that otherwise would be (more) harmful to other ecosystems. A typical example is the use of wetlands for wastewater treatment.

2. Ecosystems are imitated or copied to reduce or solve a pollution problem, leading to constructed ecosystems. Examples are fishponds and constructed wetlands for treatment of wastewater or diffuse pollution sources.

3. The recovery of ecosystems after significant disturbances. Examples are coal mine reclamation and restoration of lakes and rivers.

4. The use of ecosystems for the benefit of humankind without destroying the ecological balance, that is, utilization of ecosystems on an ecologically sound basis. Typical examples are the use of integrated agriculture and development of organic agriculture. This type of ecotechnology finds wide application in the ecological management of renewable resources.

Note that all four types use an ecological toolbox.

The idea behind these four classes of ecotechnology is illustrated in Figure 12.1. Note that ecotechnology operates in the environment and the ecosystems. It is here that ecological engineering has application of its toolbox.

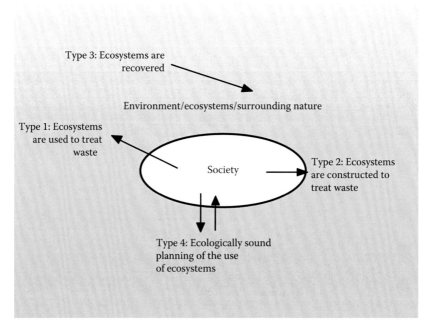

FIGURE 12.1
An illustration of the four types of ecological engineering.

Illustrative examples of all four classes of ecological engineering may be found in situations where ecological engineering is applied to replace environmental engineering because the ecological engineering methods frequently offer an ecologically more acceptable solution and ecological engineering is often the only method that can offer a proper solution to the problem. Examples are shown in Table 12.1, where the alternative environmental technological solution is also indicated. This does not imply that ecological engineering consequently should replace environmental engineering. On the contrary, the two technologies should work hand in hand to solve environmental management problems better than they could do alone. This is illustrated in Figure 12.2, where control of lake eutrophication has required both ecological engineering and environmental technology. The same combination may be used if the lake is contaminated by toxic substances as wastewater treatment of industrial wastewater is able to remove toxic substances from this source of contamination, while a wetland can be used to remove toxic substances from inflowing river water and siphoning of the hypolimnetic is able to remove toxic substance that is released from contaminated sediment. Further details are given in the third section of the chapter.

Type 1 ecological engineering, application of ecosystems to reduce or solve pollution problems, may be illustrated by wetlands utilized to reduce the diffuse nutrient loadings of lakes. This problem could not be solved by

TABLE 12.1

Ecological Engineering Examples

| Type of Ecological Engineering | Example of Ecological Engineering | | Environmental Engineering Alternative |
	Without Environmental Engineering Alternative	With Environmental Engineering Alternative	
1	Wetlands utilized to reduce diffuse pollution	Sludge disposal on agricultural land	Sludge incineration
2	Constructed wetland to reduce diffuse pollution	Root zone plant	Traditional wastewater treatment
3	Recovery of lakes	Recovery of contaminated land *in situ*	Treatment of soil by physical–chemical methods
4	Agroforestry	Ecologically sound planning of harvest rates of resources	

Note: Alternative environmental engineering methods are given, if possible.

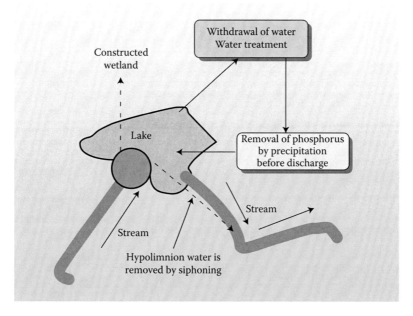

FIGURE 12.2

(**See color insert.**) Control of lake eutrophication with a combination of chemical precipitation for phosphorus removal from wastewater (environmental technology), a wetland to remove nutrients from the inflow (ecotechnology, type 1 or 2), and siphoning off of hypolimnetic water, rich in nutrients and downstream (ecotechnology, type 3). The same combination of methods could, in principle, be applied to control the toxic substance contamination of a lake because precipitation can be used to remove toxic substances from wastewater, wetland can be used to reduce the toxic substance concentration of the inflowing stream, and siphoning can be used to remove toxic substances from the hypolimnetic water released from the sediment.

environmental technology, which does not offer an applicable solution to diffuse pollution problems, including pesticides in agricultural drainage water. Treatment of sludge could be solved by environmental technology, namely, by incineration, but the ecological engineering solution, sludge disposal on agricultural land, which involves a utilization of the organic material and nutrients in the sludge, is a considerably more sound method from ecological perspectives, but of course it requires that the sludge does not contain toxic substances in harmful concentrations. The application of constructed wetlands to cope with the diffuse pollution is a good example of ecological engineering type 2. Again, this problem cannot be solved by environmental technology. The application of root zone plants for treatment of small amounts of wastewater is an example of ecological engineering, type 2, where the environmental engineering alternative, a mechanical–biological–chemical treatment cannot compete in most cases, because it will have excessive costs relative to the amount of wastewater (sewage system, pumping stations, and so on). A solution requiring less resources will always be ecologically more sound.

Recovery of land contaminated by toxic chemicals is possible using environmental technology, but it will most often require transportation of soil to a soil treatment plant, where biological biodegradation or extraction of the contaminants takes place. Ecological engineering is able to offer a treatment *in situ* with adapted microorganisms or plants. This method will be much more cost effective and the pollution related to the transport of soil will be omitted. Restoration of lakes by biomanipulation, installation of an impoundment, sediment removal or coverage, siphoning off hypolimnetic water, rich in nutrients, downstream, or several other proposed ecological engineering techniques are type 3 examples of ecological engineering. It is hardly possible to obtain the same results by environmental engineering because this requires activities in the lake and/or the vicinity of the lake.

Type 4 of ecological engineering is to a great extent based on the prevention of pollution by utilization of ecosystems on an ecologically sound basis. It is hardly possible to find environmental engineering alternatives in this case, but it is clear that a prudent resource-balanced harvest rate of renewable resources, whether of timber or fish, is the best long-term strategy from an ecological and economic point of view. An ecologically sound planning of the landscape is another example of the use of type 4 ecological engineering.

Application of biomanipulation mentioned above is somehow complex (structural changes take place; see Jørgensen, 2012), because aquatic ecosystems have two possible structures: below about 50 µg/L, where zooplankton and carnivorous fish are dominant and above 150 µg/L, where phytoplankton and planktivorous fish are dominant (see Figure 12.3). Between 50 and 150 µg/L, both ecological structures are possible, dependent on the history of the aquatic ecosystem. Biomanipulation (de Bernardi and Giussani, 1995) can be used successfully in this range—and only in this range—to make a "short cut" by removal of planktivorous fish and release carnivorous fish. If

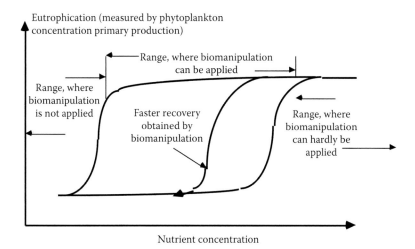

FIGURE 12.3
Hysteresis relation between nutrient level and eutrophication measured by phytoplankton concentration. The possible effect of biomanipulation is shown. An effect of biomanipulation can only be expected in the range approximately 50–150 μg P/L. Biomanipulation can hardly be applied successfully above 130–150 μg/L (see also de Bernardi and Giussani, 1995; Jørgensen and de Bernardi, 1998).

biomanipulation is used above 150 μg P/L, some intermediate improvement of the water quality will usually be observed, but the lake will later get the ecological structure corresponding to the high phosphorus concentration, that is, a structure controlled by phytoplankton and planktivorous fish. Biomanipulation is a relatively cheap and effective method provided that it is applied in the phosphorus range where two ecological structures are possible. De Bernardi and Giussani (1995) give a comprehensive presentation of various aspects of biomanipulation. Simultaneously, biomanipulation makes it possible to maintain relatively high biodiversity, which does change the stability of the system. A higher biodiversity gives the ecosystem a greater ability to meet future, unforeseen changes without changes in the ecosystem function.

By a survey of restoration methods, it is possible to conclude that ecological engineering methods often offer a cost-moderate restoration of aquatic ecosystem. The application of the methods should, however, in many cases work hand in hand with other methods, reducing the nutrient loadings, which is mainly possible by environmental technology. If, for instance, a lake has a phosphorus concentration of let us say 250 μg/L, it would probably be beneficial to combine environmental technology—treatment of wastewater to reduce considerably the phosphorus concentration in the treated wastewater—with ecological engineering—for instance, biomanipulation *when* the phosphorus concentration has been reduced to below 150 μg/L.

Illustration 12.1

A lake with a total phosphorus concentration of 0.2 mg/L is very eutrophic. 75% of the phosphorus is coming from wastewater and the remaining 25% from other sources. Give a solution to the problem, where environmental technology and ecotechnology are working together.

Solution

Since the phosphorus concentration is too high for application of biomanipulation, it would be beneficial to remove at least 90% of the phosphorus from the wastewater by chemical precipitation, which would be able to reduce the total phosphorus concentration to less than 0.1 mg/L and combine this treatment with biomanipulation. Of course, the retention time of the water in the lake determines how long time it will take to reduce the total phosphorus concentration to about 0.125 mg/L.

12.3 Agricultural Waste and Drainage Water

Particular animal waste causes great problems in intensive farming. The productions of chickens, pigs, and cattle are in many industrialized countries concentrated in rather large units, which imply that the waste from such production units requires hundreds of hectares for suitable distribution and feasible use of its value as fertilizer. Animal waste has a high nitrogen concentration, 5%–10% based on dry matter (2%–4% dry matter). If the nitrogen is not used as fertilizer, it may

1. Either evaporate as ammonia, which is partly toxic
2. Be lost to deeper layers, where it contaminates the groundwater
3. Be lost by surface runoff to lakes and streams

In addition to the high concentration of nutrients in animal waste, it may contain medicine residues and pesticides—often meaning highly toxic and harmful compounds. Agricultural waste has therefore become a crucial pollution problem in many countries with intensive agriculture. The following possibilities give a summary of the available methods, based on ecological engineering or interface between environmental technology and ecotechnology, to reduce the pollution originating from agricultural waste.

1. Storage capacity for animal waste to avoid spreading on bare fields.
2. Green fields in winter to ensure use of the fertilizing value of agricultural waste.

3. Composting of agricultural waste to assure conditioning before it is used. The composting heat may be utilized. It can be considered an intermediate method between ecological engineering and environmental engineering.

4. Anaerobic treatment of agricultural waste for production of biogas and conditioning before use as fertilizer. The biogas can be used to heat the farm buildings. It is preferable to mix cattle and swine manure to obtain the highest efficiency with this process, as an excessive free ammonia concentration (more than 1.1 g/L) results in a decreased rate of gas formation. This method may also be considered an intermediate method between ecological engineering and environmental engineering and it may be considered to belong to ecotechnic (see Section 12.1).

5. Chemical precipitation of agricultural (animal) waste with activated benthonite is applied to bind ammonium and obtain a solid concentration of 6%–10%. This process gives two advantages:

   a. The required storage capacity is reduced by a factor 2–4 (the solid concentration is increased from 2%–4% to 6%–10%).

   b. The loss of nitrogen by evaporation of ammonia is reduced by a factor of 3–8 corresponding to an adsorption of about 60%–90% of the ammonium on the added benthonite.

6. In China, animal waste is applied in fish ponds. Zooplankton eats the detritus and fish feed on the zooplankton.

Drainage water from agriculture may contain, in addition to nutrients that cause eutrophication, pesticides, and veterinary medicine residues. The best solution to this problem is the use of wetlands as protective zones between agriculture and streams, lakes, and ponds. As shown in Figure 12.4, wetlands are able to remove all these substances.

12.4 Wastewater Treatment by Ecotechnology

Traditionally, waste stabilization ponds (WSPs) are built as flow-through systems with an anaerobic, a facultative, and one or more maturation ponds. The following processes are utilized in the pond system: settling (mainly in the first ponds), anaerobic decomposition of organic matter (mainly in the first ponds), aerobic decomposition of organic matter (mainly in the last ponds, where algae are present and produce oxygen), uptake of phosphorus and nitrogen by algae (facultative and maturation ponds), evaporation of ammonia (mainly where pH is high, that is, in the last ponds), settling of algae, and denitrification (in the anaerobic zones). High removal efficiencies

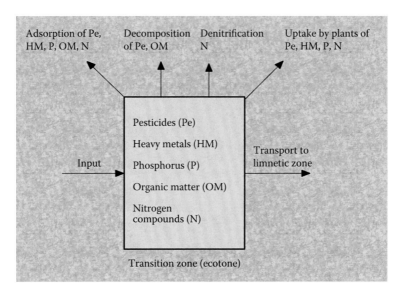

FIGURE 12.4
(**See color insert.**) Figure showing that pesticides, heavy metals, phosphorus, organic matter, including veterinary medicine residues, and nitrogen compounds can be removed by wetlands. It is therefore very important to have a wetland area as a buffer zone between agriculture and ecosystems.

of BOD5, COD, microorganisms, nitrogen, and phosphorus may be obtained provided that the guidelines for design and maintenance are followed. The method has mainly been applied in the tropical region where the temperature ensures a high rate of the processes, mentioned above.

One of the major maintenance problems is associated with removal of sludge. It is absolutely necessary to remove sludge at least twice a year because the sludge reduces the volume particularly of the anaerobic ponds and causes an undesired internal loading. The major toxic substances present in the wastewater will accumulate in the sludge, a factor that has to be considered when deciding where the sludge should be placed after removal from the WSP. Generally, an increase in the sediment depth of 0.2 cm per month can be expected. Ammonia volatilization is the main mechanism of nitrogen removal. Removal of phosphorus by WSPs is in the range of 20%–50% but depends highly on the possibility of removing algae from the effluent before discharge. Enhanced removal of phosphorus, which is often required for discharge of wastewater into lakes and reservoirs can be achieved by addition of precipitants such as calcium compounds and clay minerals, which would also increase the removal efficiency of heavy metals and some organic toxic compounds. It is recommended to examine the possibilities to use addition of cheap local and natural compounds. Addition of the chemicals to the influent may increase the removal efficiency of phosphorus and toxic compounds to about 80%–95%. Overall high efficiency is, however, best obtained in most cases by the use of a constructed wetland after the WSPs (see below), preferably with harvest of the wetland 2–4 times per year. If major amounts of toxic substances are removed by WSPs, for instance, by use of precipitation, the sludge can of course not be used as compost or soil conditioner, but should be either treated or deposited on a site where it can be controlled so that it does not contaminate the environment.

The possibilities of removing toxic substances in the constructed wetlands are illustrated in Figure 12.4. The adsorption capacity of wetlands offers significant protection against pollution by toxic substances, both heavy metals and toxic organic substances, including pesticides originating from agriculture. The ratio of the concentration of heavy metals or pesticides in organic matter to the concentration in the water of the wetland at equilibrium is strongly dependent on the composition of the organic matter, but is usually between 50 and 5000, which indicates that a wetland has an enormous binding capacity for these pollutants.

Construction of artificial wetlands is an attractive and cost-effective solution to solve the pollution problems of diffuse sources and wastewater. Wetlands are first of all able to cope with the nitrogen and heavy metal pollution from these sources, but it is essential to make a proper planning on where to place the artificial wetlands, as their effects are dependent on the hydrology (they should be covered by water most of the year and should have sufficient retention time to allow them to solve the key pollution problems) and on the landscape pattern (they should protect the most vulnerable ecosystems, which often are lakes and reservoirs). It is furthermore important to ensure that the wetlands are not releasing other components such as phosphorus as mentioned above. The following emergent types of plants are proposed for use in constructed wetlands: cattails, bulrush, reeds, rushes, papyrus, and sedges. Submerged species can be applied in deepwater zones. Species that have been used for this purpose include coon tail or horn wart, redhead grass, widgeon grass, wild celery, and watermill foil. It should also be kept in mind that a wetland in most cases will reduce the water budget due to evapotranspiration (Mitsch and Gosselink, 2007). Wetlands, however, reduce the wind speed at the water surface and may thereby also reduce the evaporation. It is important to consider these factors in the planning of artificial wetlands. Finally, it should not be forgotten that an artificial wetland is not fully developed overnight. In most cases, it will require 2–4 years for an artificial wetland to obtain sufficient plant coverage and biodiversity to be fully operational. It is, however, clear from the experience gained in the relatively few constructed wetlands that the application of models for the wetland, encompassing all the processes reviewed above, as well as for the lake is mandatory, if positive results are to be expected.

Wetlands encompassing the so-called root zone plants or subsurface wetlands may be utilized directly as wastewater treatment facilities. This application of soft technology seems particularly advantageous for developing countries due to its moderate cost. The self-purification ability of wetlands has found wide application as a wastewater treatment method in several developing countries (China, Philippines, Burma, India, and Thailand). Generally, the constructed wetlands in the tropic regions show higher efficiency than in the temperate regions due to higher photosynthetic and microbiological activity. In addition, the seasonal variations of the removal capacity of constructed wetlands are in the tropic regions much smaller than in the temperate regions. Constructed wetlands in the northern part of Europe have in the winter period (December–February) a capacity expressed, for instance, as kg N or kg COD removed/ha 24 h, which may be less than 30% of the summer capacity. This implies that constructed wetlands in the temperate zone have to be constructed with an overcapacity relative to the average situation to be able to meet the required efficiency in the winter period. These differences between constructed wetlands in tropic regions and in temperate regions explain why it is usually more expensive to use constructed wetlands in the temperate zone.

Wastewater treatment using floating species has also been proposed. Different types of duck weed and water hyacinths (*Eichhornia crassipes*) have been applied as an alternative to waste stabilization ponds. Use of water hyacinths requires, however, strict control, as they easily become widely spread as a weed, which may get completely out of control. The inorganic nitrogen and phosphorus brought in by sewage and decomposed from organic pollutants by microorganisms are absorbed by water hyacinths, which can also adsorb toxic organic compounds and heavy metals.

12.5 Soil Remediation

Today, there is a wide range of available methods to recover soil contaminated by various toxic substances. We distinguish between methods based on removal of the contaminated soil and *in situ* treatment methods. The first group of methods are usually more costly, as they entail relatively expensive transportation which in itself produces pollution. These methods are partially environmental technological methods, but they are included in this chapter (see also Section 11.4) to present a more clear overview, as they may in some cases be considered partially ecological engineering methods or ecotechnic methods. From an ecological point of view, *in situ* methods must be preferred, but they cannot be used in all possible situations.

12.5.1 Removal and Treatment of Contaminated Soil

Treatment facilities to handle contaminated soil have been constructed in many industrialized countries where the problem is most relevant. They are based on one of the three principal treatment options: incineration, bioremediation by a composting-like process, or extraction.

Incineration or thermal treatment of the contaminated soil can be used to eliminate organic contaminants susceptible to destruction. This process is energy intensive and expensive, as the transportation costs add to the incineration costs. The method is not applicable to inorganic contaminated soil, for instance, soil contaminated by heavy metals.

Biological degradation of organic contaminants is a cheaper alternative to incineration. Dedicated bioreactors may be used or just open composting sites. The latter possibility is of course more cost effective, but entails that the drainage water must be treated as it will inevitably contain the contaminants in an unacceptably high concentration. It is sometimes necessary to add water to the process to ensure rapid biological decomposition of the organic contaminants, particularly in dry seasons or arid climates. If the contaminant is difficult to decompose, because it has a very slow biodegradation rate (see Section 4.6), addition of adapted microorganisms may offer a solution.

Extraction may also be used to reclaim soil, but the method has several technical and ecological disadvantages. The extraction liquid must be treated and recovered, which is sometimes technically difficult to accomplish with high efficiency. If even a minor emission of the extraction liquid takes place, unless water is used, it will cause an additional pollution problem to be solved. Only if the contaminant can be recovered or is soluble in water, will this method offer some perspectives.

12.5.2 *In Situ* Treatment of Contaminated Soil

These methods seem more attractive from both an economic and an ecological point of view because the elimination of the transportation step entails that the cost and the pollution associated with the transportation are eliminated too. These methods are definitely ecological engineering. In addition, *in situ* treatment has the advantage that the mixture of contaminants is known, which makes it possible to tailor a treatment method to this specific mixture. If the contaminants are water soluble, it may be advantageous to dissolve them and thereby remove them from the soil. Addition of surfactants to the water may increase the solubility of the pollutants. Above the water table, extraction is normally conducted by creating a vacuum to remove vapor—the contaminants in volatilized form—from the unsaturated zone. This method has proven to be an especially useful process for the hazardous components of gasoline and other hydrocarbons. Heavy metal ions may be removed by the use of a solution of EDTA or similar ligands that considerably enhance the solubility of heavy metal ions. As the solubility of heavy metal ions in EDTA solutions is highly dependent on pH, the pH is adjusted to the optimum value (for most heavy metals around pH = 5.0).

The most desirable method of all treatment processes is *in situ* biodegradation to render the soil harmless. The rate of recovery with these processes should of course be sufficiently high to prevent the contaminants from traveling far off-site, where they might pose a risk to humans or ecosystems, before they are decomposed. A promising possibility is to apply addition of air, which is pumped down pipes to below the water table to accelerate the decomposition in the saturated and unsaturated zone. *In situ* biodegradation involves a conflict between biodegradation and mobility, which obviously is dependent on the polluting compounds. The addition of nutrients or oxygen may provide the enhancement needed to achieve acceptable rates of degradation. Also, the use of adapted microorganisms may accelerate the decomposition. The presence of easily degradable compounds can lead to cometabolism and the destruction of other refractory compounds. This is the case with high-molecular-weight PAHs in the presence of 2–3 ring polycyclic aromatics.

Recovery of soil by phytoremediation is a recently developed technique. It is clearly an ecotechnological method. Heavy metals may also be removed by application of electrolysis. This process is applicable entirely to high concentrations, which in addition makes a recovery of the metals possible. A combination of electrolysis and plant bioaccumulators seems to be an attractive solution for soil that is highly contaminated with heavy metals.

Recently, it has been shown that a few plant species (for instance, the alpine pennycress) can be used as bioaccumulators (hyperaccumulators). When they grow in contaminated soil, they are able to assimilate many times higher (10–100 times) heavy metal concentration than normal plant species. If the plants are harvested, the corresponding amount of heavy metal is removed from the soil. If the soil is only slightly contaminated with heavy metal, this method may be able to remediate the soil in a few years, which makes the method very attractive due to its low costs. For very contaminated soil, the number of years needed for complete recovery of the soil will probably be too high. By watering the plants carefully with a diluted EDTA solution, it is possible to enhance the uptake of heavy metals in the plants. Jørgensen (2009) and Mitsch and Jørgensen (2003) report a removal of 11.5% lead per harvest of plants watered by 0.02 M EDTA solution at an initial concentration of 380 mg Pb/kg soil (dry matter). Several plant species release chelating ligands and enzymes into the soil and can thereby accelerate the removal rate. Chelating agents will also reduce the toxicity of heavy metal ions (see also Chapter 7), where methods to calculate the equilibrium concentrations are presented.

Hyperaccumulators are usually plants that are slow to grow and therefore slow to remove heavy metals as mg/m^2 year. Fast-growing poplar trees are not such good hyperaccumulators as many other species, but due to their fast growth, they are promising phytoremediators.

Aquatic plants *Salvinia* and *Spriodela* have been used to remove chromium and nickel from wastewater. A significant removal efficiency in the concentration range 1–8 ppm, although with fluctuations, has been reported (Mitsch and Jørgensen, 2003). Fungal biomass has recently been proposed for the removal of heavy metals. The amount of metal adsorbed per unit weight is several mg/g, for cadmium, copper, lead, and nickel. The plants can efficiently take up organic substances that are moderately hydrophobic, with a K_{ow} from about 0.5 to 3.0. More hydrophobic substances exceeding a K_{ow} (the ratio of the solubility in octanol and water) of 3.0 approximately bind so strongly to the soil and the roots that they are not easily taken up within the plants.

References

De Bernardi, R. and G. Giussani (editors). 1995. *Biomanipulation in Lakes and Reservoirs Management*, ILEC. UNEP, Otsu, Nairobi, 208pp.

Jørgensen, S.E. (editor). 2009. *Applications in Ecological Engineering*. Elsevier, Amsterdam, 380pp.

Jørgensen, S.E. 2012. *Introduction to Systems Ecology*. CRC Press, Boca Raton, 320pp., Chinese edition (2013).

Jørgensen, S.E. and R. de Bernardi. 1998. The use of structurally dynamic models to explain the success and failure of biomanipulation. *Hydrobiologia* 379: 147–158.

Luca Palmeri, L., A. Barausse, and S.E. Jørgensen. 2013. *Ecological Processes Handbook*. CRC Press, Boca Raton, 386pp.

Mitsch, W.J. and L.G. Gosselink. 2007. *Wetlands*, 4th edition. Van Nordstrand Reinhold, New York, 580pp.

Mitsch, W. J. and S.E. Jørgensen (editors). 1989. *Ecological Engineering: An Introduction to Ecotechnology*. John Wiley and Sons, New York, p. 430.

Mitsch, W. J. and S.E. Jørgensen. 2003. *Ecological Engineering and Ecosystem Restoration*. John Wiley and Sons, New York, 412pp.

Odum, H.T. 1962. Man in the ecosystem. In: *Proceedings of the Lockwood Conference on the Suburban Forest and Ecology*, Bull. Conn. Agr. Station 652, Storrs, Connecticut, pp. 57–75.

Odum, H.T. 1983. *Systems Ecology*. Wiley Interscience, New York, 520pp.

Straskraba, M. 1985. *Simulation Models as Tools in Ecotechnology Systems: Analysis and Simulation*, Vol. II. Academic Verlag, Berlin, pp. 362–372.

Straskraba, M. 1993. Ecotechnology as a new means for environmental management. *Ecological Engineering* 2: 311–332.

13

Application of Cleaner Production in Environmental Management

13.1 Introduction

All human activities cause an impact on the environment. Particularly, agriculture, industry, and their products together with the production processes and following waste disposal have caused many pollution problems. We can, in principle, solve these problems by environmental technology and ecotechnology. This, however, obviously raises the question: Could we not change the production, the production method, or even the products in such a way that we would be able to reduce the impact on the environment—pollute less but get the same results, so to speak? Such a change in production, for instance, by recycling inside the production unit may be even much cheaper to realize than the treatment of the subsequent emissions. It may even give a cost reduction, as recycling and an increased efficiency in the use of the resources obviously means less use of "virgin" materials or resources. It is sometimes a little erroneously called green audit, but environmental audit is more often used today. It is to a great extent a question about modeling the mass and energy flows of the production. There are numerous examples of simultaneous reductions in costs and environmental impact due to changes in a production method. At the same time, it is interesting to note that it was earlier believed that it would only pay to implement cleaner technology and production on major production units or factories but experience has shown that even smaller- and medium-sized enterprise units may benefit from such considerations (Klewitz and Hansen, 2014).

The various technological approaches (and environmental legislation) can work hand in hand with the basic idea to find the optimum solution based on the diagnosis by playing on the entire spectrum of available methods. The improvement of production processes with considerations on how to save the burden on the environment is widely known today under the name of *cleaner production*. In this context, it should be realized that concerns of the environment not only deals with the relations of a factory to its externals but also relates to its employees by considering the internal environment. In fact, improving production by reducing the load on the surrounding environment takes on various forms that may be organized in a hierarchical manner. At the lowest level, small improvements of the efficiencies of the use of basic resources such as water and electricity could be mentioned. The next level can be exemplified by hard-core technological consideration of changing the production process and addressing even the design of the product. The highest level could be a basic innovative step, sometimes called *industrial ecology* where several production units are (re)organized or restructured for the benefit of the environment. Both the areas of cleaner production and industrial ecology have today their own dedicated journals and both serve the purpose of increasing the sustainability of necessary productions of our society (Almeida et al., 2015; Korhonen, 2004).

It is difficult to find the optimum solution to an environmental problem because of the many possible solutions including those based on combinations of methods, which often in practice may give the optimum solution. Not only technological but also social and organizational issues such as LEAN-ing activities may be taken into consideration (Medeiros et al., 2014; Neto and Jabbour, 2010). This also means that a concerted action in the area is bound to be multidisciplinary. In many industries, it has been close to the optimum solution to enhance recycling, reduce water consumption, and to introduce many minor changes in production, which would together at least reduce emissions.

Improved environmental management of productions can be achieved mainly by four initiatives:

1. Introduction of an environmental management system (EMS) or better an environmental management and audit scheme (EMAS) (EU, 2009, 2013) requires a rather detailed knowledge about the mass and energy flows of a factory or production site, which unfortunately is not always readily available. Thus, the first step would be to acquire this information and examine all the production steps from an environmental view point: could we not reduce the emission here and there? How do we do it? This step often leads to environmental certification (e.g., EMAS) of the industry and, for instance, the use of ecolabeling (Hale, 1996), which may be utilized in negotiation with contractors and customers. Many global and regional firms require that their subsuppliers have environmental certification in order to ensure that their final products are "green." The following ISO standards support the enterprises in their effort to improve their green image and reduce emissions by internal changes in production: ISO 14,001–14,004 and partly ISO 140,014. This is the topic of the next section.

2. The establishment of environmental auditing requires that the enterprise set up a balance for energy usage and mass of all

components. The quantities of consumed raw materials, produced products, and material lost as pollutants to the environment at various points of production must be indicated. Such knowledge may have been gained through the establishment of life cycle inventories, the so-called LCIs (e.g., Weidema and Wesnæs, 1996) or full life cycle assessment (LCA) of products (Azapagic, 1999; Reap et al., 2008a,b). Such studies investigate the environmental impacts from cradle to grave or even from cradle to cradle (Braugart and McDonough, 2008). The latter approach raises attention and focuses on the possibilities of a change in design of products and even a total redesign with the purpose of closing cycles further by reusing and in this context touching upon the ISO 9000 standards. The approaches typically use what is known as a *functional unit* such as the production of a certain mass of product or waste as a base for evaluation of the environmental load of a specific production. The ISO standards 14,010–14,013 give guidance for environmental auditing. This is the topic of Section 13.3.

3. The implementation of cleaner technology and production covers the actual changes in the production frequently resulting from 1 and 2. It is an attempt to provide a full answer to the pertinent question: how can we produce our products by a more environmentally friendly method? Which changes in the production should be made? The ISO standards mentioned under 1 and 2 are supplied in ISO 14,014, 14,015, and 14,031. Section 13.4 is devoted to a cleaner technology. Clearly, these examinations reveal the use of all raw materials. If toxic substances are used or substances that may harm the working and external environment, it is obviously questioned, whether these substances could be replaced by less toxic or less harmful substances. It would, of course, require a good overview of the properties of all applicable chemicals and compounds and an answer to these relevant concerns that environmental chemistry and ecotoxicology are widely applied—not only in the chemical industry but in the entire industry, as a large number of chemical compounds and substances are used industry today. Remember in this context, as it has been mentioned a few times previously in this book, that the industrialized countries apply about 100,000 chemicals today.

4. Not only the production method but also the product should be considered to be changed, for instance, by changing design and change of raw materials and as indicated by point 3 all chemicals used in the production. Environmental risk assessment, introduced in Chapter 6, may be used to change from a substance with a high environmental risk to a product with a low environmental risk. A life cycle analysis, LCA, examines all the emissions associated with the use of the product from the cradle to the grave. The idea is to examine the

possibilities for reductions in emissions by minor or major changes in the product, including possibilities of recycling the final discarded waste product. This topic will be discussed further in Section 13.5.

The cleaner production methods discussed and presented in this chapter are not all associated with toxic substances but they are presented because they may inspire to be used also for the cases where toxic substances are involved. In principle, a cleaner production includes of course the out phasing of toxic substances. The particular focus on toxic substances and the possibilities to replace them or change chemical production to produce less toxic waste is presented in Section 13.6. The last section is devoted to the Reach Reform, which illustrates how environmental legislation can be used in addition to environmental technology, ecotechnology, and cleaner technology. It has of course been widely discussed whether the Reach Reform is sufficiently green.

13.2 Environmental Management Systems

The regulation, control, and requirements of documentation of the impact of industrial activities on health and the environment has undergone a significant transition since the late 1970s and early 1990s in the EU, North America, and Japan. Since the mid-1980s, industries themselves have taken a more proactive stance in recognizing that sound environmental management on a voluntary basis can enhance corporate image, increase profits, reduce costs, and obviate the need for further environmental legislation in the area. The commercial values of green products have been fully acknowledged. In some countries, the government has even set up what may be named a "green purchase policy," which is to be followed not only by the state but also by counties and municipalities, fiscal authorities, and governmental institutions. It is therefore not surprising that there has been an enormous interest among many enterprises. Introduction of an EMS can help companies to approach environmental issues systematically in order to integrate environmental care as a normal part of their operation and business strategy. The major triggers for the companies to initiate such environmental changes are rooted in legislation, in pressure from the stakeholders, in the importance of image and reputation, in the competition (environmental certification is required for subsuppliers and the commercial value of a green product), and in economies (sales potential, recycling means lower consumption of resources, green taxes).

ISO 14,001 uses the following definition of the term EMS: that part of the overall management system which includes organizational structure, planning activities, responsibility, practices, procedures, processes, and resources for developing, implementing, achieving, reviewing, and maintaining the environmental policy.

EMS is intended to

- Identify and control the environmental aspects, impacts, and risks relevant to the organization
- Develop and achieve its environmental policy, objectives, and targets, including compliance with environmental legislation
- Establish short-, medium-, and long-term goals for environmental performance, ensuring a balance of costs and benefits, for the organization and its stakeholders
- Determine what resources are needed to achieve these goals, assign responsibility for them, and commit the necessary resources
- Define and document specific tasks, responsibilities, authorities, and procedures to ensure that every employee acts in the course of his or her daily work to help achieve the defined goals
- Communicate these throughout the organization and train people to effectively fulfill their responsibilities effectively
- Measure performance against preagreed standards and goals and modify the approach if necessary

EMS has benefited mainly from two developments:

A. The rising cost of environmental liabilities led companies to develop environmental auditing as a management tool.
B. Total quality management concepts were adopted to reduce defects = noncompliance with specifications.

The development of EMS may be described in several phases (Dyndgaard and Kryger, 1999):

1. Set up an *environmental policy*, expressing the commitments of the enterprise to specific environmental goals. These are most clear if they are *quantified*, for instance, to reduce the emission of a specific pollutant by 90%. This point also includes decisions on disclosure of production and results of LCA's and corporate social responsibility (CSR) issues and initiatives.
2. An environmental *program of action plan* describing the measures to be taken over a given period of time. The action plan should identify the steps to be taken by each department to reduce emissions, commit the necessary resources, and provide monitoring and coordination of progress toward the achievement of the defined goals.
3. *Organizational structures* delegating authority and assigning responsibility for actions. In many cases, a whole new structure has to be built within the organization.

4. *Integration* of environmental management into business operations and internal communication including risk assessment to identify potential accidents.

5. *Monitoring* and measurements and record-keeping procedures to document and monitor the results as well as the overall effects of environmental improvements.

6. *Corrective actions* to eliminate potential nonconformances to objectives, targets, and specifications.

7. EMS *audits* to check the adequacy of efficiency of the implementation.

8. *Management reviews* to assess the status and adequacy of the EMS in light of changing circumstances.

9. *Internal information and training* to ensure that all employees understand to fulfill their environmental responsibility and why they should.

10. *External communications* and community relations to communicate the enterprise's environmental goals and performance to interested persons outside the enterprise to obtain the advantage of a green image.

All 10 steps are covered in detail by ISO 14,001 and are consistent with the quality management in ISO 9000, which discusses the four phases: the planning phase, action phase, evaluation phase, and correction phase. Performance measures are needed in EMS to provide indication of how well the goals have been reached and to serve as an incentive for the employers as their efforts are recognized and appreciated through the results. The reward to the enterprise can be substantial because even a limited EMS can often demonstrate that the economic gains can outweigh the costs and that inexpensive measures may yield important environmental and financial returns.

The effect of EMS introduced into different enterprises including several industries with high pollution potential (manufacturers of dye stuff, cables, enzymes, and printed matter) have been examined. The results of the examination are summarized in the following list of advantages including indirect advantages gained by introduction of EMS:

1. A direct measurable reduction of emissions with all the benefits that may entail.

2. A general increased engagement by all employees not only with EMS but also with all the enterprise's areas of activities.

3. Improved (green) image.

4. The economic gains outweigh all environmental investments and associated costs from within a few months to a maximum of 3.8 years.

5. In several cases, new orders have been obtained directly due to the introduction of EMS.

6. Internal reductions in noise, dust, and odors were obtained.

7. Optimization of processes and better utilization of resources.

8. Better cooperation with environmental agencies.

13.3 Environmental Audit, LCI, and LCA

In order to implement an EMS or EMAS, it is necessary to create an overview of the inputs and outputs of the production system—to set up mass and energy balances. This can be done by establishing a green accounting system or environmental audit covering a mass and energy balance of the raw materials and energy used in a production unit. When the production is rather complex, it may be beneficial to develop a model of the mass and energy cycling in the production. Through a quantification of the mass and energy flows in a production unit (e.g., a factory), often presented in the form of one or more flowcharts, it becomes possible to identify how much materials and energy are lost and where this takes place. Hence, the established balances are at the crux of LCIs and LCAs. These results enable us to answer the following pertinent questions:

- Could the losses (=emissions) be reduced? Which are quantitatively and qualitatively significant? What may be achieved immediately by implementation of already existing technology?

- Could the loss of energy eventually be recycled? How would that influence the other processes of the production? If the production scheme is complex, a model will be very helpful to answer these questions.

- How would these changes resulting from the answers of the two above-mentioned questions influence the mass and energy balances and the quality of the product?

- How could the mass and energy flows be optimized with relation to the total costs, including energy, raw material, environmental, labor, and miscellaneous expenditures?

Mass and energy balances are in this context often called *green accountings* or *environmental audits*, as they are comparable with accounts of income (useful outputs), expenses (inputs), profit (value of useful outputs – inputs), and loss (nonuseful outputs). An account is settled by looking at the money flow in the same manner, using the principles of conservation: the input – the output corresponds to the accumulation in the firm = an increased asset.

When toxic substances are involved, it implies that the use, the inclusion in the products, and the loss of toxic substances are determined. This does not require that the environmental audit be made by an economic auditor which has been the tendency, but it is preferable that it is carried out by engineers who know the production and the possibilities to change the processes and thereby reduce the environmental problems. Mass balances, particularly in the chemical industry, are often based on stoichiometric calculations that require a good knowledge of chemistry (environmental chemistry) and the involved chemical technological processes. Similarly, energy balances can hardly be set up without the use of thermodynamics and studies of energy transfer processes. The environmental issues are, however, in most environmental audits covered at least partially, while 8%–10% of the audits give very incomplete information about the environmental impact. This examination has also uncovered the economy behind the use of green audit. In several cases, the enterprises show gains from the results of environmental audit.

It is often surprising how little enterprises know about their losses (emissions to the environment) before they introduce a green audit. Most enterprises focus almost entirely on the two P's: product (quality and price) and profit. By introduction of a green audit many enterprises have, however, acknowledged that they can increase profit by reduction of the losses (emissions) and decreasing the price and increasing the quality of the product from a commercial point of view by developing a green image. A few examples are presented below to illustrate how a green audit has been able to elucidate the value of waste.

Fibrous material is lost to the wastewater in paper and pulp production. During the 1970s, it was realized on the basis of mass balance that a significant amount of fibrous material was lost to the environment. It is, however, possible using chemical precipitation with benthonite and a polyflocculant as precipitant to collect the fibers as sludge (5%–10% dry matter), which can be used to produce lower quality paper.

Mass balances have also revealed that 25–40 years ago the amounts of fish oil lost by fish filleting plants and the amounts of chromium (chromium(III)) lost by tanneries were very significant particularly from an economic point of view. Today, a fish filleting plant that is not recovering fish oil or a tannery that is not recycling chromium cannot survive the tight competition.

Generally, it can be strongly recommended to develop mass and energy balances for all productions because it will frequently reveal where it is possible to reduce the losses and gain efficiency in the use of matter and energy.

13.4 Cleaner Production/Technology

Cleaner production is in accordance with the United Nations Environment Programme (UNEP) and is defined as "the continuous application of an integrated preventative environmental strategy to processes, products, and services so as to increase efficiency and reduce the risks to humans and the environment." The main emphasis is on all aspects of the manufacturing processes and products but now the attention has been drawn to the importance of resource efficiency. Cleaner production therefore includes the efficient use of raw materials and energy, the elimination of toxic and dangerous materials, and the reduction of emissions and wastes at the source. In addition, it focuses on reduction of the impacts along the entire life cycle of the products and services—from design to use and to ultimate disposal (for further details; see Section 13.5). Table 13.1 shows a comparison of pollution control based mainly on environmental technology and cleaner production based on all the approaches presented in this chapter: EMS, environmental audit, changes of processes and technologies applied in production, and LCA.

Cleaner production is especially important to developing countries (it also explains UNEP's interest in this topic), because it provides industries in these countries for the first time with an opportunity to bypass the more established industries that are still stocked with the costly environmental technology to solve the problems at the end of the pipeline. Table 13.2, mainly based on Dyndgaard and Kryger (1999) and partially on the examples above in the context of environmental audit, gives several examples of how sometimes relatively simple changes or introduction of low-cost equipment can provide cleaner production with reduced emissions and lower costs. Note how fast the investment in the change to cleaner production is paid back in some cases.

TABLE 13.1

Comparison of Pollution Control Approaches and Cleaner Production Approaches

Pollution Control Approaches	Cleaner Production Approaches
Pollutants are controlled with filters and treatment methods	Pollutants are prevented at their sources wastewater through integrated measures
Pollution control is evaluated when processes and products have been developed and when the problems arise	Pollution prevention is an integrated part of product and process development
Pollution control is considered to be a cost factor	Pollutants and wastes are considered to be potential resources
Environmental challenges are to be addressed by environmental experts	Environmental challenges are the responsibility of people throughout the company
Environmental improvements are to be accomplished with technology	Environmental improvements include nontechnical and technical approaches
Environmental improvements should fulfill authorized standards	Environmental improvements involve a process of working continuously to achieve higher standards
Quality is defined as meeting the customer's requirements	Quality means the products that meet the customer's needs and have minimal impacts on human health and the environment

TABLE 13.2

Examples of Cleaner Production

Industry	Country	Method	Waste and Emission Reduction	Pay Back Period
Fish filleting	Denmark	Removal of fish oil from wastewater	75% reduction of BOD5	3 months
Tannery	Denmark	Recycling of chromium	95% reduction of chromium in wastewater	<1 year
Paper	Switzerland	Precipitation of fibers	70% reduction of BOD5	2–3 years
Fruit juice	The Philippines	Collection of fruit drops	55 L organic waste/h	9 months
Wood finishing	Malaysia	Waste segregation	54,000 kg of hazardous waste + 5700 m^3 w.w.	3 months
Lead oxide	India	New insulation in furnace	Fuel consumption reduced 50%, increase production 3%	3 months
Cement	Indonesia	Control system to ensure optimum temperature	3% reduced energy use, 9% higher production, reduced emission of NO$_x$	<1 year

Kryger and Dyndgaard (1999) attempted to review the barriers to introduction of cleaner production, including bureaucratic resistance, conservatism, uncoordinated legislation, misinformation, scarce money, and lack of technical reliable information.

ILLUSTRATION 13.1

Compare a solution of an environmental problem based on (1) environmental technology, (2) ecological engineering, and (3) cleaner technology/production considering (a) investment, (b) running costs, (c) environmental aspects (solution of the pollution problem), and (d) ecological aspects (impacts on ecosystems).

Solution

	1	2	3
(a)	High	Medium	Medium
(b)	High	Low	Low to medium
(c)	Good	Good but may f(time)	Good
(d)	Good, depends on the ecological aspects; are included in the solution	Usually good	Good, depends on the ecological; aspects are included

The focus for (1) and (3) is often only on the pollution problems and not on the ecological consequences, which explains the answers in the table.

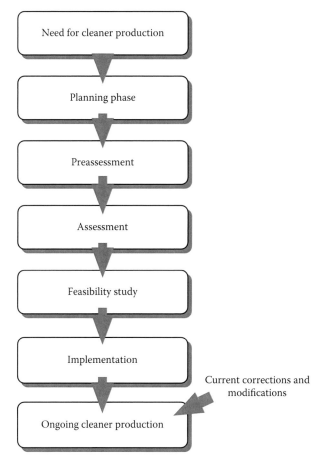

FIGURE 13.1
Introduction of cleaner production by a seven-step procedure.

Environmental technology is usually more expensive to apply as well as the investment costs and the running costs. Because nature has a high variability, the use of ecotechnology will inevitably have a high variation in the efficiency as a function of time.

Introduction of cleaner production often follows the procedure presented in Figure 13.1. An option generating process is often used extensively from the planning stage to the implementation. It is often beneficial to consider the following elements in this process (Kryger and Dyndgaard, 1999):

1. *Change in raw materials* either by changing to another (more pure) grade of the same raw material, eventually by purification of the raw material or by substituting the present raw material with a less hazardous material.

   Example: Detergents containing polyphosphates are replaced with detergents based on zeolites.

2. *Technological changes* covering changes in the production pro-cesses, modification of equipment, use of automation, and changes of process conditions (flow rates, temperatures, pressures, resi-dence times, and so on). The aim with these changes is to reduce the emissions and/or improve the quality of the product.

Example: Use of mechanical cleaning to replace cleaning by chemicals.

3. *Good operating practices* can often be implemented with a little cost. They include management and personnel practices, mate-rial handling and inventory practices, training of employees, loss prevention, waste segregation, cost accounting practice, and production scheduling. Example: Reduce loss due to leaks.

4. *Product changes* include changes in quality standards, product composition, and product durability or product substitution.

Example: Replace chlorofluorocarbons (CFCs) with ammo-nia in refrigerators.

5. *Reuse and recycling* involves the return of a waste material to the originating or another process to substitute input material.

Examples: Recover nickel plating solution with an ion exchanger, recover dye stuffs from wastewater in the textile industry by ultrafiltration.

These five possibilities for cleaner production are summarized in Figure 13.2.

The range of tools to catalyze industry to adopt cleaner production is large (Kryger and Dyndgaard, 1999). Governmental agencies may use the following instruments:

A. Regulations, standards, legislations
B. Green taxes
C. Economic support
D. Providing external assistance

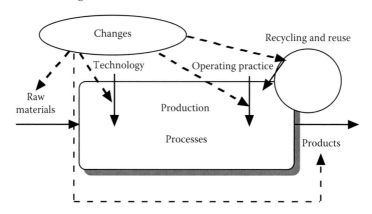

FIGURE 13.2
Cleaner production can be introduced by changes (dotted arrows) in raw materials, technol-ogy, operating practice, products, and reuse or recycling.

The two principles to solve pollution problems, A (to reduce emissions by cleaner production) and B (reuse and recycling), work hand in hand and can hardly be separated. B is, however, not always a consequence of cleaner technology but may be an obvious solution after environmental technology has been introduced.

On the other hand, cleaner technology/production may apply environmental technology as a tool to achieve the goals set up in a cleaner production scheme. This illustrates very clearly that *integrated* environmental management is needed to achieve an optimum solution to a pollution problem. To date, integrated environmental management requires that the entire spectrum of considerations and tools presented in this volume is applied.

13.5 Life Cycle Inventory and Life Cycle Assessment

LCA is a methodology for assessing the environmental impacts and resource consumption associated with a product throughout its entire life—from cradle to grave or from raw material to production to product to use of product to disposal (Braugart and McDonough, 2008). It is a holistic approach, using system analysis. Several industrialized countries are discussing or have already implemented LCA. The following initiatives have enhanced the interest by industry for introduction of LCA:

1. Ecolabeling or environmental declarations about products based on LCA
2. Green purchase guidance taking environmental considerations into account
3. Take-back responsibility for certain products, for example, cars and electronics

LCA consists of four steps (Hautschild and Wenzel, 1999):

1. *Definition of the goals of the assessment and the scope of the study.* The goals are often formulated as questions:

   Which product on the market causes the least impact on the environment?

   Which are the most important environmental impacts of the product during its life cycle? What would happen if the product was not produced at all?

   How does the impact of a specific product during its life cycle compare with alternative products (comparative studies which should be based on functional units)?

What influence would a specific proposed change in the product or production processes have on the results of the LCA?

Which properties of the product cause the undesirable impacts? Can we change them? How?

2. *Preparation of an inventory of inputs and outputs* for the processes that occur during the life cycle of the product. Traditionally, emissions are divided into air pollutants, wastewater, and solid waste. The emissions are expressed relative to the function of the product in order to make a proper comparison.

3. *Impact assessment*, where the results under point 2 are translated into environmental impacts. We distinguish the impact on human health, ecosystem health, and the resource base. Various techniques are used to compare two or more emissions with the same or different types of effects. When the emissions have the same type of effects, a normalization may be applied, while different effects require the introduction of weighting factors. The selection of weighting factors is at least to a certain extent a political question, as it is impossible by objective criteria to compare effects on human health and on ecosystems or the depletion of two different resources. Table 13.3 shows an example of the proposed weighting factors, which of course are open for a discussion.

4. *Interpretation of the impact assessment according to the defined goals and scope of the study* including sensitivity analysis of key elements of the assessment. This phase attempts to answer the questions raised in the first phase. The outcome may be a series of recommendations to the decision maker, but may also lead to revision of the scope of the study. The result of the sensitivity analysis is used to identify which product properties and processes are most significant for the total result. The significance of the uncertainty of the key figures

TABLE 13.3

Weighting Factors Proposed by Wenzel et al. (1997)

Impact	Weighting Factor
Global warming	1.3
Ozone depletion	23
Photochemical smog formation	1.2
Acidification	1.3
Nutrient enrichment	1.2
Persistent toxicants	2.5
Human toxicity	2.8
Ecotoxicity	2.3

is examined by letting them vary within their estimated range and investigating how this affects the total result of the LCA and the conclusions that can be drawn.

The following stages of the life cycle are considered in a complete LCA:

A. Extraction of raw materials.
B. Transportation of raw materials to the production site.
C. Manufacture of the product.
D. Transportation from production to the user. This may consist of several stages.
E. Use of product.
F. Disposal.

In all these stages, emissions may occur and the following treatment processes may be considered as the most obvious alternatives: reuse (after a suitable treatment), recycling, incineration, composting, wastewater treatment, and dumping. The result of the LCA is changed in accordance with any decision on treatment of the waste.

13.6 Green Chemistry

Obviously, the chemical industry shows a particular endeavor to apply cleaner technology because the spectrum of chemicals used is very wide from completely natural and harmless chemicals to very hazardous substances. Furthermore, there are wide possibilities because of the high number of chemicals to replace toxic substances by more harmless chemicals. Anastas and Warner (1998) have proposed 12 principles of green chemistry, which are completely consistent with that of the cleaner production approach presented in Table 13.1:

1. Prevention is environmentally better and in most cases more cost moderate than treating and cleaning up waste after it has been created.

2. Synthetic methods should be designed to maximize incorporation of all material used in the process into the final product, whereby the amount of waste products are minimized.

3. Substances that possess little or no toxicity to human health and environment should be used in the design of the production and the product.

4. Chemical products should be designed to affect their desired function while minimizing their toxicity.

5. Auxiliary substances (solvents and separation agents) should be made unnecessary wherever possible. If it is not possible, alternatives to volatile organic solvents, chlorinated solvents, and environmentally harmful solvents should be chosen.

6. Energy requirements should be recognized and minimized and the efficiency in the use of energy should be maximized.

7. Raw material should be renewable whenever technically and economically feasible.

8. Unnecessary derivatization should be minimized or avoided.

9. Catalytic reagents are often superior to stoichiometric reagents.

10. Chemical products should be designed to be degradable and do not persist in the environment.

11. Analytical methodologies need to be developed for real-time, in process monitoring, and control prior to the formation of hazardous substances.

12. Substances and chemical processes should be chosen to minimize the potential for chemical accidents (releases, explosion, and fire).

Sustainability, industrial ecology, ecoefficiency, and green chemistry direct the development of next-generation materials, products and processes, exemplified by a current trend to replace oil-based plastic by bio-based polymers. Generally, with advances in biotechnology, these possibilities are increasing. Biocatalysis and biosynthesis (Ran et al., 2008) have been a significant accomplishment in the use of biologically based raw materials as a viable alternative to petroleum-based materials. The increasing use of bio-based energy sources is supported by these trends:

- Biogas produced from waste and manure
- Use of vegetable oil as diesel fuel in cars
- Production of diesel fuel from micro- and macroalgae
- Production of bioethanol from corn or even better second generation of bioethanol production based on straw, garbage, and other waste material
- Use of fast-growing plants and trees (for instance, willow trees) as biofuel

A further rapid growth of green chemistry is expected in the coming years due to an enormous innovative potential in catalytic chemistry, a wide spectrum of new biological compounds and processes, use of nanotechnology, and ultrasound technology.

13.7 Reach Reform

In EU law, a premarket testing of chemicals was introduced in 1980. This requirement of premarket testing was very reasonable, but a problem was that the chemicals already in the market had not been tested, at least not properly, and that very often their properties were not uncovered sufficiently to be able to develop an applicable ERA: What about these chemicals? The obligations of testing existing substances were, however, not imposed on the industry but on public authorities. The major bulk of chemicals were introduced before 1980 and therefore remain untested, because it would require a very long time and be very costly to perform the testing of this major bulk of chemicals. In addition, the chemicals introduced after 1980 had to be tested, which of course was a disadvantage for the new chemicals and posed a hindrance for innovation of more environment friendly and better chemicals. The Reach Reform was adapted in 2006 to try to solve these problems. The main elements of this reform are as follows:

1. A substance must be registered before it can be put on the market, if used in quantities above 1 t/year. Substance manufacturers and importers into EU are obliged to send the registration with safety data and potential uses to the ECHA (European Chemicals Agency).
2. A chemical safety report is required by ECHA if the manufacturers and importers use 10 t or more of the substance per year. The report should include a hazard and risk assessment with an exposure scenario of the substance for the specified uses.
3. ECHA reviews the registration and may demand additional information. ECHA will, however, not have the resources to review the quality and accuracy of all registrations.
4. When the registration process is complete, the manufacturers and importers produce/import the substance within the limits of what is indicated in the registration.
5. ECHA will check the compliance of a minimum of 5% of the registrations.
6. ECHA can in cooperation with EU member state authorities request further information on particular substances when there are suspicions on risk for humans, animals, and environment. The costs for these procedures should be shared among registrants of the substance.
7. Evaluation may lead to the conclusion that the use of certain substances must be put under restrictions or that authorization is considered necessary to some substances.

8. For substances of very high concern, the authorization is required for their use and placing on the market. These substances are considered to have hazardous properties of such high concern that it is necessary to regulate them centrally.

There are several deficiencies in Reach Reform, but it must still be considered as a step forward in the control with the use of chemical substances and it has served as an example for other agencies outside EU.

References

Almeida, C.M.V.B., F. Agostinho, B.F. Gianetti, and D. Huising. 2015. Integrating cleaner production into sustainability strategies: An introduction to this special volume. *Journal of Cleaner Production* 96: 1–9.

Anastas, P.T. and J.C. Warner. 1998. *Green Chemistry: Theory and Practice*, Oxford University Press, Oxford.

Azapagic, A. 1999. Life cycle assessment and its application to process selection, design and optimisation. *Chemical Engineering Journal* 73: 1–21.

Braugart, M. and W. McDonough. 2008. *Cradle to Cradle: Remaking the Way We Make Things.* North Point Press, New York, 230 pp.

Dyndgaard, R. and J. Kryger. 1999. Environmental management system. In S.E. Jørgensen (ed.), *A Systems Approach to the Environmental Analysis of Pollution Minimization.* Lewis Publication, Boca Raton, pp. 67–87.

EU. 2009. *Regulation (EC) No 1221/2009 of the European Parliament and of the Council of 25 November 2009 on the Voluntary Participation in a Community Eco-Management and Audit Scheme (EMAS).*

EU. 2013. *Commission Decision of 4 March 2013 Establishing the User's Guide Setting Out the Steps Needed to Participate in EMAS, under Regulation (EC) No 1221/2009 of the European Parliament and of the Council on the Voluntary Participation by Organisations in a Community Eco-Management and Audit Scheme (EMAS).*

Hale, M. 1996. Ecolabelling and cleaner production: Principles, problems, education and training in relation to the adoption of environmentally sound production processes. *Journal of Cleaner Production* 4 (2): 85–95.

Hauschild, M. and H. Wenzel. 1999. LCA. In: S.E. Jørgensen (ed.), *A Systems Approach to the Environmental Analysis of Pollution Minimization.* Lewis Publication, Boca Raton, pp. 155–191.

Jørgensen, S.E. 2006. *Eco-Exergy as Sustainability.* WIT, Southampton. 220pp.

Klewitz, J. and E.G. Hansen. 2014. Sustainability-oriented innovation of SMEs: A systematic review. *Journal of Cleaner Production* 65: 57–75.

Korhonen, J. 2004. Industrial ecology in the strategic sustainable development model: Strategic applications of industrial ecology. *Journal of Cleaner Production* 12: 809–823.

Kryger J. and R. Dyndgaard. 1999. Introduction to cleaner production. In S.E. Jørgensen (ed.), *A Systems Approach to the Environmental Analysis of Pollution Minimization.* Lewis Publishing, Boca Raton, pp. 87–101.

Medeiros, J.F., J.L.D. Ribeiro, and M. Nogueira. 2014. Sucess factors for environmentally sustainable product innovation: A systematic literature review. *Journal of Cleaner Production* 65: 76–86.

Neto, A.S. and C.J.C. Jabbour. 2010. Guidelines for improving the adoption of cleaner production in companies through attention to non-technical factors: A literature review. *African Journal of Business Management* 4 (19): 4217–4299.

Ran, N., L. Zhao, Z. Chen, and J. Tao. 2008. Recent applications of biocatalysis in developing green chemistry for chemical syntheses at the industrial scale. *Green Chemistry*. 1: 361–372.

Reap, J., F. Roman, S. Duncan, and B. Bras. 2008a. A survey of unresolved problems in life cycle assessment. Part 1: Goal and scope and inventory analysis. *International Journal of Life Cycle Assessment* 13: 290–300.

Reap, J., F. Roman, S. Duncan, and B. Bras. 2008b. A survey of unresolved problems in life cycle assessment. Part 1: Impact assessment and interpretation. *International Journal of Life Cycle Assessment* 13: 374–388.

Weidema, B.P. and M.S. Wesnæs. 1996. Data quality management for life cycle inventories—An example of using data quality indicators. *Journal of Cleaner Production* 4 (3–4): 167–174.

Wenzel, H. et al. 1997. *Environmental Assessment of Products. Volume 1. Methodology. Tools and Case Studies in Product Development*. Chapman & Hall, London. 378pp.

14

Implementation of Integrated Environmental Management to Solve Toxic Substance Pollution Problems

14.1 Introduction

Holistic, integrated environmental management must consider all the seven steps presented in Chapter 1. Simultaneously or expressed differently, all the toolboxes must be applied and integrated into a concerted effort to be able to go from a complex environmental problem to the solution. We need to integrate the entire spectrum of knowledge about the problem and its sources and effects and the knowledge gained by use of the diagnostic tools, which will lead us to a composite and combined application of the solution toolboxes. The goal is to solve the problems, including the side problems and all derived problems, proximate as well as distal. It is important in the process of solving a focal environmental problem not to create new and eventually overlooked environmental problems. It is important to overview all the possible environmental problems simultaneously. It is recommended to list all the possible environmental problems at an early stage and consider how these problems are interrelated and how an integrated environmental management could consider an overall (holistic) solution.

It is of course a difficult task to keep in mind all the seven steps and all the toolboxes in the development of, if not the very best solution, at least a fully acceptable and appropriate solution. It is clearly necessary to maintain a holistic view and overview of the environment and its reactions and processes originating from the problem. Integrated ecological and environmental management means that the environmental problems are viewed from a holistic angle considering the ecosystem as an entity and considering the entire spectrum of solutions, including all possible combinations of proposed solutions. The integration in this context means that all the available tools and possibilities are taken into account and all the problems are considered simultaneously.

The experience gained from environmental management in the last 40 years has clearly shown that it is important not to consider solutions of single problems but to consider *all or at least all major* problems associated with a considered ecosystem simultaneously and evaluate *all* the solution possibilities proposed by the relevant disciplines at the same time, or expressed differently: to observe the forest and not the single trees. The experience has clearly underlined that there is no alternative to an *integrated* management, at least not on a long-term basis. Fortunately, we have seen in the previous chapters that new ecological subdisciplines have emerged and they offer toolboxes to perform an integrated ecological and environmental management, provided we are using a "macroscope" meaning applying a holistic integrated view. Following all the seven steps (see Chapter 1) will facilitate the macroscopic view.

The use of ecological modeling has become much more important over the last 40 years, due to a need for quantification in ecology and environmental management. This is probably the only significant tool available for obtaining a fully quantitative, holistic overview of the environmental problems associated with complex ecosystems, such as grasslands, forests, savannas, lakes, rivers, wetlands, lagoons, or reservoirs, which is a prerequisite for selecting an optimum solution to the complex problems facing them. Because solutions to the entire spectrum of environmental problems require quantitative estimation methods to assist in identifying a realistic tradeoff between ecological and economic concerns, it is not surprising that ecological models have been used increasingly in environmental management efforts. Ecological indicators and assessment of ecosystem services are more recent tools than ecological modeling, but can supplement ecological modeling with slightly different views. They can work hand in hand with ecological models but can only partially replace ecological modeling in some cases even if a completely quantitative understanding of the problem and the possible solutions is needed, but they can usually be applied with less effort and therefore less costs than ecological modeling. Many ecotoxicological models have been developed during the last 50 years, and use of indicators based on concentrations or effects of toxic substances in the environment have had a wide and useful application, too. Loss of ecosystem services due to the emission of toxic substances has only been applied in a few cases, because the use of ecosystem services as diagnostic tool has been introduced more recently.

We may assume that it is most often beneficial to apply all three diagnostic tools to achieve the widest overview, such as in the case of a pollution problem associated with toxic substances. Nevertheless, the final decisions about the implementation of the diagnostic tools are eventually an economic issue, and certainly a problem facing environmental agencies is how to efficiently distribute funds to enhance natural capital and improve ecosystem services (e.g., Hajkowicz et al., 2009; Pinto et al., 2014). In fact, according to Mysiak et al. (2002), "Environmental decisions require a spatial analysis of the impacts resulting from possible options in order to guarantee spatially balanced sustainable development." Hence, in the processes of sustainability assessments and/or environmental planning of ecosystems, it becomes fundamental to ensure the accurate interweaving of disciplinary approaches, and as already said, the three diagnostic tools must be considered for this task. The use of several approaches is always advised to achieve a better evaluation of ecosystem integrity combined with current human demands, perceived as key for the sustainable development of ecosystems.

In fact, sustainable development is a complex issue, involving strongly interconnected ecological, social, and economic aspects of an ecosystem, now and in the future (de Jonge et al., 2012). Eventually, according to Kiker et al. (2005), "Decision-making in environmental projects can be complex and seemingly intractable, principally because of the inherent tradeoffs between socio-political, environmental, ecological, and economic factors." Decisions affect different stakeholder groups in different ways, leading to inevitable tradeoffs between them and between present and future generations. Decision making involving such complex systems requires a logically well-structured process.

In this scope, multicriteria analyses (MCA) have been recommended as useful tools to ensure an integrated management of an ecosystem, allowing the incorporation of different sets of data (e.g., de Jonge et al., 2012; Pavlikakis and Tsihrintzis, 2003; Villa et al., 2002). MCA constitutes a stepwise process that allows comparing of decision alternatives with multiple and often complex impacts (Hermann et al., 2007). The information is often structured using a software tool, which aims to record proposed alternatives, while measuring and assessing their impacts (Hermann et al., 2007).

14.2 Recommendations on Application of Integrated Environmental Management

From this discussion, several important recommendations may be deduced, as summarized in the following six points:

- Expect that a proper environmental strategy will require a wide spectrum of approaches and techniques, which to a high extent must be based on the presented diagnostic tools.
- Expect that a proper environmental management will require the application of a combination of end-of-the-pipe technologies (environmental technology), ecotechnology, cleaner technology, and environmental legislation. In this context remember that toxic substances, as all emissions, have several sources and that they participate in a number of environmental pathways and processes, which make a proper abatement strategy very complex.

- Correct timing in applying the various steps in environmental management efforts is extremely important; thus, it is recommended that a comprehensive environmental management plan be developed at a very early stage, in order to use the available resources in the most optimal manner.

- It is usually very beneficial, particularly from an economic perspective, to consider prevention, rather than corrective adjustments, primarily because it is often very costly to restore heavily degraded ecosystems, if restoration is even possible. This is recommended for all environmental problems, but particularly for problems associated with toxic substances, because damage caused by toxic substances is often irreversible, interfering with human health, and more expensive to eliminate sufficiently. There are numerous cases that illustrate that the harms and damages by toxic substances are significant and irreversible and that careful considerations before the application of the toxic substances often emphasize that prevention is far the best method and even in some cases the only applicable method.

- Because of the complexity of ecosystems and their problems, proper ecological knowledge about ecosystems is a prerequisite for ecologically sound environmental management programs. This is the only reasonable method for avoiding unexpected ecosystem responses; therefore, all the ecological subdisciplines must be implemented in the effort to find a proper solution.

- Optimum solutions to environmental management problems are best obtained if all the ecosystems of the entire watershed/landscape are taken into consideration in developing and implementing management actions to avoid the possibility that the solution of the environmental problems of one ecosystem creates other problems in other ecosystems. Remember in this context that all ecosystems are open systems and that toxic substances may use many processes and pathways in the environment.

Principles and quantifications are the keywords in our search for solutions to environmental problems. Chapter 1 has clearly presented how to go from

1. Problems to
2. Which ecosystem is involved to
3. Sources (with an appropriate estimation of the quantities) to understand the impacts and a determination of the effects that the impacts have on the ecosystem to
4. A diagnosis, using the well-developed diagnostic tools to
5. Selection among the possible solutions to the topic of this chapter to

6. Integration of these steps into an environmentally sound plan, using a suitable combination of the methods found in point 5, and based on the knowledge obtained in each step, to
7. Following the recovery

The environmental plan will, by following the seven steps, inevitably be based on the answers to the following questions (see Jørgensen, 2000; Jørgensen and Nielsen, 2012):

A. What can be obtained by changing the methods that are applied today to produce the products needed by the society to build and construct, to use products, matter, and energy? The answer builds very much in its widest sense on cleaner technology and production. Section 13.6 gives several examples and discusses particularly this solution possibility.

B. Could we use beneficially—both economically and environmentally—the three Rs, reduce, reuse, and recycle? If yes, it points toward a wide use of cleaner technology, which has particular possibilities to combine what we call green chemical solutions with the three Rs, see Section 13.6.

C. Would it be an advantage to decompose waste causing environmental problems to harmless, or at least less harmful, components that in some cases could even be reused or recycled but under all circumstances could be discharged without additional harmful effects, for instance, decomposing a toxic substance to harmless compounds?

D. Is there a solution that removes waste to another location where the deposition would be more harmless or may be even beneficial? The answer to this question requires a good insight into the alternative location and the ecological and other effects on this location. This solution should be applied very carefully when the environmental problem is a toxic substance. It is, however, applied, for instance, for a final deposition of radioactive waste.

E. Can the deteriorated ecosystem be recovered by ecotechnological methods? It is particularly attractive to use for contaminated soil (see Section 12.5), but it is also applicable for aquatic ecosystems.

F. Can the problem be avoided by a better ecological planning, which could include the use of green chemistry.

Figure 14.1 (Jørgensen, 2000) illustrates how a life cycle analysis can be applied to integrate the various solution toolboxes, considering the entire life cycle for a product from its production to its final disposal. The figure illustrates well, therefore, the idea behind integrated environmental management. The arrows to the ecosystems presume that the ecological diagnostic

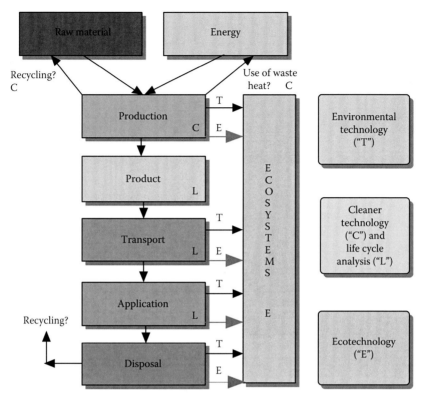

FIGURE 14.1
(**See color insert.**) The arrows cover mass flows and the thin arrow indicates the control possibilities. Point pollution is indicated by a thin black arrow to the ecosystems and nonpoint pollution is indicated by a thick gray arrow. Production includes both industrial and agricultural production. The latter is most responsible for nonpoint pollution. Environmental technology (indicated by T) is mostly applied to solve the point pollution problems and ecotechnology (indicated by E). Cleaner technology (indicated by C) is investigating the possibilities to produce the products by another method that causes less environmental problems and may facilitate reuse and recycling. The diagram is based on a life cycle analysis—the product is followed from its production based on matter and energy to its final disposal as waste after its use. The diagram shows how the various solution tool bases are needed in an integrated way to solve all the problems associated with a product from its production to its disposal.

toolboxes are applied and the sources indicated as the arrows are quantified and their impacts are known.

14.3 Following the Recovery Process

The focal environmental problem is not solved before examining that the expected recovery of the involved ecosystems has taken place. It requires that the damages of the ecosystems have been eliminated and that the concentrations of the toxic substances and their derived substances have been reduced to the level that is in accordance with the environmental management plan. A monitoring program to assess these recovery criteria is needed. The program should be an integrated part of the entire environmental management, and it is important that the results of the monitoring program is made accessible to the public to enable all citizens to evaluate the success of the environmental management. It is important that everybody living in the region can answer the question: Has the recovery been completed?

The monitoring program is most often quite obvious in the sense that the previous steps of the integrated environmental management have clearly defined the goals of the recovery process, both with respect to the elimination of ecosystem damages and levels of toxic substances as a function of time and space.

References

de Jonge, V.N., R. Pinto, and K. Turner. 2012. Integrating ecological, economic and social aspects to generate useful management information under the EU Directives' "Ecosystem Approach." *Ocean Coast Manage.* 68: 169–188.

Hajkowicz, S., A. Higgins, C. Miller, and O. Marinoni. 2009. Is getting a conservation model used more important than getting it accurate? *Biol. Conserv.* 142: 699–700.

Hermann, B.G., C. Kroeze, and W. Jawjit. 2007. Assessing environmental performance by combining life cycle assessment, multi-criteria analysis and environmental performance indicators. *J. Clean Prod.* 15: 1787–1796.

Jørgensen, S.E. 2000. *Principles of Pollution Abatement.* Elsevier, Amsterdam, 520pp.

Jørgensen, S.E. and S.N. Nielsen. 2012. Tool boxes for an integrated ecological and environmental management. *Ecol. Indic.* 20: 104–109.

Kiker, G.A., T.S. Bridges, A. Varghese, T.P. Seager, and I. Linkov. 2005. Application of multicriteria decision analysis in environmental decision making. *Integr. Environ. Assess. Manage.* 1(2): 95–108.

Mysiak, J., C. Giupponi, and A. Fassio. 2002. Decision support for water resource management: An application example of the MULINO DSS. *Conference IEMSS 2002—Integrated Assessment and Decision Support*, IEMSS, Lugano, pp. 138–143.

Pavlikakis, G.E. and V.A. Tsihrintzis. 2003. A quantitative model for accounting human opinion, preferences and perceptions in ecosystem management. *J. Environ. Manage.* 68(2): 193–205.

Pinto, R., M.C. Cunha, C. Roseta-Palma, and J.C. Marques. 2014. Mainstreaming sustainable decision-making for ecosystems: Integrating ecological and socio-economic targets within a decision support system. *Environ. Process.* (DOI 10.1007/s40710-014-0006-x).

Villa, F., L. Tunesi, and T. Agardy. 2002. Zoning marine protected areas through spatial multi-criteria analysis: The case of the Asinara Island National Marine Reserve of Italy. *Conserv. Biol.* 16(2): 515–526.

15

Summary and Conclusions

15.1 Feasibility of Implementing Integrated Environmental Management

Jørgensen et al. (2015) and Jørgensen and Nielsen (2012) have presented how it is possible to carry out integrated environmental management in seven steps. The integration is based on the application of several available toolboxes that are able to

1. Identify the environmental problem both qualitatively and quantitatively
2. Develop a diagnosis that relates the environmental problem and its effects and consequences
3. Combine several toolboxes that offer solutions to the problem, to suggest a proper ecological–economical solution

It has been central to the development of an integrated, holistic environmental management that a number of ecological subdisciplines have been developed in the last 45 years: ecological modeling, ecological engineering, application of ecological indicators, and assessment of ecosystem services and systems ecology, which are a basis discipline of the other subdisciplines.

The seven steps and the idea behind integrated environmental management are presented in the first chapter. In Chapter 1, the problems of applying integrated environmental management as presented by Jørgensen et al. (2015) have been briefly mentioned. However, this book attempts to answer the question: Is it feasible to apply the proposed integrated holistic environmental management on the environmental problems associated with our application of about 100,000 chemicals in our society?

I do hope that the readers agree with me that the answer to this pertinent question is "yes." The seven steps and the toolboxes can be applied to solve environmental, chemical, and ecotoxicological problems. There are, however, some differences from other environmental problems. The focal environmental problems of this book are associated with the enormous amount of detailed knowledge about the chemicals and their effects that we need to implement the seven steps and the toolboxes. It informs that we are forced to use estimation methods to a great extent, and although a number of estimation methods are available, the application of estimation implies a higher uncertainty or standard deviation.

To compensate for the relatively high uncertainty (standard deviation), a high assessment factor is applied. The assessment factor should of course be applied to the benefit of the environment, not the economy, to ensure that the environmental problems are solved. Only in a few cases has it been difficult to use sufficiently high assessment factors. It cannot, of course, be tolerated that human health or fundamental properties of nature on which we are very dependent are threatened. Therefore, high assessment factors are applied when our knowledge about the effects of chemicals is very limited (see the discussion in Chapter 8 on ERA). As discussed in Chapter 9, the general use of the assessment factor implies that it is acceptable in most cases to use relatively simple (uncomplicated) models to develop a good diagnostic analysis. The uncertainty of the model needs to be less than the uncertainty expressed by the assessment factor.

15.2 Conclusions

It has been demonstrated throughout the book that the application of the procedure (Jørgensen et al., 2015) for an integrated, holistic environmental management in seven steps can be beneficially implemented for environmental problems associated with our extensive use of more or less toxic chemical substances. Let us look at the seven steps and comment on the conclusions that we can make based on the 14 previous chapters of the book:

1. *Define the problem*: Definition is of course an answer to the pertinent question: Which toxic substances have been released to and threaten the environment?

2. *Determine the ecosystems involved*: Obviously, the following questions have to be answered: To which ecosystems have these toxic substances been discharged? Which observations show the damages on the ecosystems and their biological, chemical, and physical components?

3. *Find and quantify all the sources to the problem*: Quantification is key information in integrated environmental management. It is a question about measuring the inflows of the toxic substances to the involved ecosystems. Chapter 2 provides an overview of the possible

sources and an extensive use of analytical chemistry is needed to determine the key concentrations.

4. *Set up a diagnosis to understand the relation between the problem and the sources*: A good diagnosis relating the amount of toxic chemical discharged to the involved ecosystems requires a good knowledge about the properties of the chemical substances, including their effects on the components of ecosystems, particularly the biological components. In some cases, we can find the needed information in the scientific literature and handbooks, but for many chemical substances we do not have the required knowledge and must use estimation methods. Chapter 3 gives an overview of the important properties of the chemicals; Chapter 4 discusses the possible estimation methods, including software package applicable for the estimations; Chapter 5 covers the global element cycles, which inevitably will influence all the chemical processes including the toxic substance processes in the ecosphere; and Chapter 6 discusses the effects of the chemicals on ecosystems and their components. Integrated environmental management requires, however, that we develop knowledge about the relation between the concentrations of the toxic substances and the effects. As all chemical substances undergo chemical processes, it is obviously necessary to examine which components will be the final results of these processes in the ecosystems. The needed calculations in this context are presented in Chapter 7, while Chapters 8 through 10 present the diagnostic tools (ERA, ecological modeling, and ecological indicators) that make it possible for the environmental managers to relate the discharge of toxic substances to the effects, including the answer to the pertinent question: How much can the effects be reduced by a given reduction of the discharge?

5. *Determine all the tools we need to implement to solve the problem*: The diagnosis will lead to an indication of how much is needed to reduce the discharge and obtain a desired, well-defined quality of the involved ecosystems. This leads to the question: Which toolboxes can we implement to obtain this reduction when a toxic substance is the issue. As for environmental management in general, we have for removal of toxic substances four toolboxes: environmental technology, ecotechnology (ecological engineering), cleaner technology, and environmental legislation. Not surprisingly, in the case that the environmental problem is a toxic substance, we have to use other tools from these four toolboxes than for environmental problems in general; see Chapters 11 through 13, where these tools are particularly emphasized.

6. *Implement the selected solutions*: For all environmental problems, a combination of the available methods often gives the best solution

with respect to both economy and ecology. Therefore, in this phase it is very important to be open to the application of a combination of methods and examine carefully the possible combinations of the applicable methods. Chapter 14 discusses these issues.

7. *Follow the recovery process*: The environmental management plan is not complete unless it can be assessed that the recovery has taken place according to the plan. Therefore, it is always necessary to follow the recovery process. For environmental problems associated with toxic substances, it is particularly important to observe that the damages that the toxic substances have made on the ecosystems have been eliminated and the lower levels of toxic substances as a function of time and space have been reached (see Chapter 14).

The overall conclusion is therefore that an integrated, holistic transdisciplinary management can indeed be applied for the environmental problems of toxic substances, provided the tools particularly aimed for these problems, presented in this book, are implemented.

References

Jørgensen, S.E., J.C. Marques, and S.N. Nielsen. 2015. *Integrated Environmental Management: A Transdisciplinary Approach*. CRC Press, Boca Raton, 380pp.
Jørgensen, S.E. and S.N. Nielsen. 2012. Tool boxes for an integrated ecological and environmental management. *Ecol. Indic.* 20:104–109.

Index

A

Absorption, 231; *see also* Air pollution
    abatement
  absorber reagents, 232
  adsorbent selection, 232–233
Acid–base reactions, 104–105; *see also*
    Chemical processes
  double logarithmic diagrams,
    112–116
Acid formers, 239
Active solid matter, 225
Activity, 101; *see also* Reactions and
    equilibrium calculations
  coefficient, 101–102
Adsorption, 31, 110; *see also* Absorption;
    Reactions and equilibrium
    calculations
  Langmuir adsorption isotherm, 110
  Lineweaver–Burk plot, 111
  log–log plot of Freundlich
    adsorption isotherm, 111
  octanol–water distribution
    coefficient, 110
  soil–water distribution, 110
Agricultural waste, 245, 254; *see also*
    Solid waste treatment
Air pollution abatement, 224; *see also*
    Absorption; Environmental
    technology
  air pollution problems, 224, 231
  carbon hydrides, 227
  catalytic oxidation, 228
  CO_2 pollution, 227
  combustion, 233
  fossil fuel, 228
  gas removal from industrial
    effluents, 231
  particulate pollution, 225–226
  SO_2 emission standards, 229
  SO_2 recovery, 230.
  sources of heavy metals, 234
  sulfuric acid, 228

Alkalinity, 117, 119; *see also*
    Reactions and equilibrium
    calculations
Allergy, 34
Allometric estimation methods, 39
Ammonia volatilization, 256
Anaerobic digestion, 238–239; *see also*
    Solid waste treatment
Anthracene, 92
Aquatic ecosystem pollution, 15; *see also*
    Pollution

B

BCF, *see* Biological concentration factor
    (BCF)
Benzo(*a*) pyrene, 92
Bioaccumulation, 24; *see also* Toxic
    substance pollution
Bioaccumulators, 260; *see also* Soil
    remediation
Biochemical recycling of matter,
    60; *see also* Global element
    cycles
Bioconcentration, 25; *see also* Toxic
    substance pollution
Biodegradability, 30
Biodegradation, 31, 259; *see also* Soil
    remediation
Biological concentration factor (BCF),
    27, 37
Biological oxygen demand (BOD), 14
Biomagnification, 25; *see also* Toxic
    substance pollution
Biomanipulation, 253
Biotechnology, 247–248; *see also*
    Ecotechnology
BOD, *see* Biological oxygen demand
    (BOD)
Buffer capacity, 116, 118; *see also*
    Reactions and equilibrium
    calculations

C

Cadmium, 84; *see also* Toxic substance
accumulation in body, 86
balance for Danish agricultural
land, 85
biomagnification and
bioaccumulation, 84
contamination of agricultural land,
85
effects, 86
uptake from food, 86
Cancer, 33
Carbamates, 90, 91
Carbonate solubility, 127–128; *see also*
Reactions and equilibrium
calculations
Carbon dioxide pollution, 227; *see also*
Air pollution abatement
Carbon hydrides, 227
Carbon-to-nitrogen ratio (C/N ratio),
241
Catalytic oxidation, 228
Central atom, 131
Centrifugation, 238; *see also* Solid waste
treatment
CFCs, *see* Chlorofluorocarbons (CFCs)
Chelation, 131
Chemical oxygen demand (COD), 14
Chemical processes, 104; *see also*
Reactions and equilibrium
calculations
acid–base reactions, 104–105
complex formation, 105–106
conditional equilibrium constant,
107
in environment, 97
precipitation and dissolution, 105
redox reactions, 106–107
Chemical risk assessment, 160–163, 164;
see also Environmental Risk
Assessment (ERA)
Chlordane, 91
Chlorine, 144
Chlorofluorocarbons (CFCs), 14
Chlorohydrocarbons, 90
Cleaner production, 8, 263; *see also*
Integrated environmental
management

environmental audit, 269–270
environmental management
systems, 266–269
examples of, 272
functional unit, 265
green chemistry, 277–278
green purchase policy, 266
illustration, 272–275
improved environmental
management of, 264–266
industrial ecology, 264
life cycle assessment, 275–277
life cycle inventory, 275
pollution control approaches vs., 271
Reach Reform, 279–280
seven-step procedure, 273
technology, 270–275
C/N ratio, *see* Carbon-to-nitrogen ratio
(C/N ratio)
COD, *see* Chemical oxygen demand
(COD)
Combustion, 233; *see also* Air pollution
abatement
Complex ecosystems, 284
Complex formation, 105–106, 131; *see
also* Chemical processes;
Reactions and equilibrium
calculations
in aquatic systems, 133
conditional constant determination,
136–139
coordination formation, 131
environmental importance of,
131–132
equilibrium calculations of, 135
redox potential and, 150–152
under simultaneous reaction, 128
triangle diagram, 133
Composting, 240–241, 242; *see also* Solid
waste treatment
Conditional constant, 133; *see also*
Reactions and equilibrium
calculations
acid–base reaction, 134
conditional equilibrium constant,
107, 134
example, 135–136
Coordination formation, 131
for metals of aquatic chemistry, 132

Corporate social responsibility (CSR), 267
CSR, *see* Corporate social responsibility
     (CSR)

D

Detergents, 93; *see also* Toxic substance
Dioxins, 88, 89; *see also* Toxic substance
Direct-acting carcinogens, 33
Dissolved carbon dioxide, 119; *see also*
     Reactions and equilibrium
     calculations
   equilibrium constants of carbon
        dioxide–carbonate, 120
   example, 123–124
Domestic garbage, 239–243; *see also*
     Solid waste treatment
Double logarithmic diagrams, 112; *see*
     also Reactions and equilibrium
     calculations
   for acid–base system, 114
   for aquatic system, 120
   conditional constant determination,
        136–139
   equations, 113
   example, 121–123
   Henderson–Hasselbach equation, 112
   pK values, 115–116
   proton balance, 113
   for redox process, 143
Dumping grounds, 240; *see also* Solid
     waste treatment

E

EC, *see* Effect concentration (EC)
Eco-exergy, 9, 209, 213; *see also*
     Ecosystem services
Ecological–economic models, 194; *see*
     also Ecological model
Ecological engineering, 6–7, 247; *see also*
     Ecotechnology; Integrated
     environmental management;
     Soil remediation
   agricultural waste and drainage
        water, 254–255
   biomanipulation, 253
   ecological modeling and, 248
   eutrophication control, 251

examples, 251
illustration, 254
nutrient level and eutrophication,
     253
toxic substance removal, 256, 257
types of, 250
Ecological indicators, 5; *see also*
     Integrated environmental
     management; Ecosystem
     health assessment (EHA);
     Ecosystem services
   characteristics of, 203
   classification, 204–206
   ecosystem health, 202
   ecosystem integrity, 202
   for EHA, 200, 203
   evolution of complex ecosystems, 202
   selection criteria, 202, 203–204
Ecological magnification factor (EMF),
     37
Ecological model, 3–6, 167, 184–186; *see*
     also Integrated environmental
     management
   calibration and validation, 181–184
   components, 171–173
   development, 168
   differential equations, 177
   ecological–economic models, 194
   ecotoxicological models vs., 187–188
   food chain dynamic models, 188–189
   groups, 186
   illustration, 173–174
   literature sources, 182
   as management tool, 169–170,
        193–197
   Michaelis–Menten equation, 178, 179
   phosphorus cycle, 175, 176
   physical and mathematical models,
        167–169
   pollution problem models, 186
   population models, 189–191
   sensitivity analysis, 180
   simple DDT model, 190
   software STELLA, 176, 177
   steady-state models of mass flows,
        189
   toxic substance in trophic level, 189,
        190
   toxic substance models, 187

Ecological modeling, 248, 284
procedure, 173, 174
process equations, 178
Ecosystem, 59, 60, 167; *see also*
Ecological indicators; Global
element cycles
health of, 202
Ecosystem health assessment (EHA),
199; *see also* Ecological
indicators; Ecosystem services
indicators for, 200
role in environmental management,
199
Ecosystem services, 206; *see also*
Ecological indicators;
Ecosystem health assessment
(EHA)
cultural services, 207–208
eco-exergy, 209, 213
provisioning services, 206–207
regulating services, 207
supporting services, 207
sustainability of nature, 208
value of, 208–213
work capacity, 211
work energy, 208–211
Ecotechnic, 247; *see also* Ecotechnology
Ecotechnology, 247; *see also* Ecological
engineering; Soil remediation
artificial wetland construction, 257
classification of, 249
phosphorus and nitrogen removal,
256
toxic substance removal in
wetlands, 256, 257
wastewater treatment, 255–258
Ecotoxicological models, 187–188; *see
also* Ecological model
with effect components, 191–192
in population dynamics, 189–191
toxin in organism vs. time, 192
Ecotoxicological properties, 33
EEP, *see* Estimation ecotoxicological
parameters (EEP)
Effect concentration (EC), 33
EHA, *see* Ecosystem health assessment
(EHA)
EIAs, *see* Environmental impact
assessments (EIAs)

Electron activity, 139; *see also* Redox
equilibria
EMAS, *see* Environmental management
and audit scheme (EMAS)
EMF, *see* Ecological magnification
factor (EMF)
EMS, *see* Environmental management
system (EMS)
Endocrine system, 34
Environmental audits, 263, 269; *see also*
Cleaner production
Environmental engineering, 248
Environmental impact assessments
(EIAs), 20, 34, 155; *see
also* Environmental Risk
Assessment (ERA)
Environmental management and audit
scheme (EMAS), 264; *see also*
Cleaner production
Environmental management system
(EMS), 264, 266; *see also*
Cleaner production
advantages, 268–269
development of, 267–268
Environmental Risk Assessment (ERA),
4, 29, 155
assessment factors to derive PNEC,
161
chemical risk assessment, 160–163,
164
concepts, 155
development of, 160–166
and EIA, 155–156
example, 159
for human exposure, 166
industrial pollutant treatment, 156
legislation and regulation, 157
risk assessment, 157
risk identification, 165
spatial and time scale for hazards
and for ecological hierarchy
levels, 158
uncertainties, 159–160
Environmental technology, 6, 215; *see
also* Air pollution abatement;
Integrated environmental
management; Solid waste
treatment; Wastewater
treatment

Equilibrium constant, 98; *see also*
      Reactions and equilibrium
      calculations
    conditional, 107
    example, 99–100
    free energy, 98
    mass law, 99
    standard free energy, 98
Estimation ecotoxicological parameters
      (EEP), 35, 40, 52–53
    allometric estimation, 39
    application, 42–52
    biodegradability of chemicals, 35–36
    biodegradation, 37–39
    connectivity by, 47
    correlation equation application,
      36–37
    estimation methods network, 41
    estimation of biological parameters,
      39–42
    octanol–water distribution
      coefficient and biological
      concentration, 38
    opening screen image of, 42
    water solubility and octanol–water
      distribution coefficient, 36
Eutrophication problem, 18, 69
Extraction, 259; *see also* Soil remediation

F

Filter belts, 238; *see also* Solid waste
      treatment
Filter presses, 238; *see also* Solid waste
      treatment
Fine dust, *see* Particulate matter
Food chain dynamic models, 188–189;
      see also Ecological model
Food web dynamic models, *see* Food
      chain dynamic models
Fossil fuel, 228
Free energy, 98
Fugacity, 20, 22
Fungicides, 90

G

Gaseous pollutants, 14; *see also*
      Pollution

Genetically modified organism (GMO),
      186
Global carbon cycle, 61; *see also* Global
      element cycles
Global element cycles, 55, 60
    atomic composition of four spheres,
      65
    biochemical recycling in
      ecosystems, 60
    biochemistry of nature, 55–57
    carbon cycle, 61
    carbon dioxide concentration vs.
      year, 62
    C, N, P as % dry matter, 56
    ecosystem, 59
    elemental composition of freshwater
      plants, 56
    element cycles and recycling in
      nature, 57–60
    nitrogen cycle, 58, 59, 62, 63
    phosphorus cycle, 57, 58, 59
    sulfur cycle, 63–64
Global nitrogen cycle, 62, 63
Global sulfur cycle, 63–64
GMO, *see* Genetically modified
      organism (GMO)
GNP, *see* Gross national product (GNP)
Green accountings, *see* Environmental
      audits
Green audit, *see* Environmental audits
Green chemistry, 277–278; *see also*
      Cleaner production
Green purchase policy, 266; *see also*
      Cleaner production
Gross national product (GNP), 17
Groundwater pollution, 15; *see also*
      Pollution

H

Hazardous concentration (HC), 33
HC, *see* Hazardous concentration (HC)
Heavy metal, 233–234; *see also* Toxic
      substance
    atmospheric pollution sources of,
      234
    pollution, 69
    in River Rhine, 78
Henry's constant, 27

Henry's law, 21, 108–110; *see also*
    Reactions and equilibrium
    calculations
Herbicides, 90
Hospital wastes, 243–244; *see also* Solid
    waste treatment
Hydroxide solubility, 124–127; *see also*
    Reactions and equilibrium
    calculations
Hyperaccumulators, *see*
    Bioaccumulators
Hypersensitivity, *see* Allergy

I

IGFCS, *see* Intergovernmental Forum
    on Chemical Safety (IGFCS)
Incineration, 241–242, 259; *see also* Soil
    remediation; Solid waste
    treatment
Industrial and mining waste, 244; *see*
    also Solid waste treatment
Industrial ecology, 264; *see also* Cleaner
    production
Informal priority setting (IPS), 157
Inorganic coagulants, 237
Inorganic compound, 67; *see also* Toxic
    substance
  concentrations and applications of, 70
  freshwater plant composition, 67
  lead in food, 71
  pollution by, 68
  toxicity as LD_{50} and LC_{50}, 71
Integrated environmental
    management, 1, 9–11, 283
  complex management, 7
  ecological engineering, 6–7
  ecological indicators, 5
  ecological models, 3–6, 284
  and ecology, 10
  implementation feasibility, 291–292
  mass flow and control possibilities,
    288
  recommendations on application of,
    285–288
  after recovery process, 288–289
  recovery process, 8–9
  steps, 1, 292–294
  sustainable development, 285

toolbox environmental technology,
    6–8
Intergovernmental Forum on Chemical
    Safety (IGFCS), 157
IPS, *see* Informal priority setting (IPS)

L

Langmuir adsorption isotherm, 110; *see*
    also Adsorption
LAS, *see* Linear alkyl benzene
    sulfonates (LAS)
LC_{50}, *see* Lethal concentration 50 (LC_{50})
LCA, *see* Life cycle assessment (LCA)
LD_{50}, *see* Lethal dose 50 (LD_{50})
Lead, 76; *see also* Toxic substance
  in aquatic ecosystems, 79
  for average European today, 80
  in Danish agriculture land, 83
  from food, 71, 79
  in glacial ice, 77
  half-life in humans, 81
  land treatment, 82
  sources of, 81
  toxicity of, 78
  toxicological effects, 80
  in water, 78
Lethal concentration 50 (LC_{50}), 33
  and LD_{50}, 71
Lethal dose 50 (LD_{50}), 33
  and LC_{50}, 71
Life cycle assessment (LCA), 265; *see*
    also Cleaner production
  steps, 275–277
Ligand, 131
Ligand atom, 131
Light pollution, 14; *see also* Pollution
Linear alkyl benzene sulfonates
    (LAS), 93

M

MAC, *see* Maximum allowable
    concentration (MAC)
Mass law, 99
Maximum allowable concentration
    (MAC), 33
Maximum permissible level (MPL), 163
MCA, *see* Multicriteria analyses (MCA)

MEA, *see* Millennium ecosystem assessment (MEA)
Mercury, 72; *see also* Toxic substance
applications, 72
distribution model, 75
in fish, 74, 75
model for Mex Bay, 74
poisoning, 72
pollution, 73, 76
Methane formers, 239
Methyl tertiary butyl ether (MTBE), 228
Microbiological decomposition, 31
Microcosmos, 168
Millennium ecosystem assessment (MEA), 206
Mixed equilibrium constant, 102; *see also* Reactions and equilibrium calculations
example, 103
mixed acidity constant, 102
Models, 168; *see also* Ecological model
Molar fraction, 116; *see also* Reactions and equilibrium calculations
MPL, *see* Maximum permissible level (MPL)
MTBE, *see* Methyl tertiary butyl ether (MTBE)
Multicriteria analyses (MCA), 285
Multidentate, 131
Mutation, 33

N

NAEL, *see* Nonadverse effect level (NAEL)
Naphthalene, 92
Narcotic concentration effect (NC), 33
NC, *see* Narcotic concentration effect (NC)
Negligible level (NL), 163
Nernst law, 140; *see also* Redox equilibria
Nitrogen cycle, 58, 59, 62, 63; *see also* Global element cycles
NL, *see* Negligible level (NL)
NOAEL, *see* Nonobserved adverse effect level (NOAEL)
Noise pollution, 14; *see also* Pollution
Nonadverse effect level (NAEL), 165

Noneffect concentration (NEC), 29, 33
Nonobserved adverse effect level (NOAEL), 165
Nygård Algae index, 205

O

Octanol–water distribution coefficient, 110
Organic compounds, 87; *see also* Toxic substance
dioxins, 88–89
PAHs, 91–92
PCBs, 88, 89
pesticides, 90–91
petroleum hydrocarbons, 87
Organometallic compounds, 92–93; *see also* Toxic substance
Organophosphates, 90, 91

P

PAHs, *see* Polycyclic aromatic hydrocarbons (PAHs)
Particulate matter, 14; *see also* Pollution
Particulate pollution, 225; *see also* Air pollution abatement
control, 226
control equipment characteristics, 226
particle size, 225
Particulate polycyclic organic matter (PPOM), 225
PCB, *see* Polychlorinated biphenyls (PCB)
PEC, *see* Predicted environmental concentration (PEC)
pe–pH diagrams, 147; *see also* Reactions and equilibrium calculations
equations for pe, 147
example, 149–150
for iron, 151
for sulfate–sulfur–hydrogen sulfide system, 149
Pesticides, 90–91; *see also* Toxic substance
Petroleum hydrocarbons, 87; *see also* Toxic substance
Phenanthrene, 92

Phosphoric acid, 116
Phosphorus cycle, 57, 58, 59, 175; *see also*
        Global element cycles
Photochemical ozone and smog, 14; *see*
        also Pollution
Phytoremediation, 260–261; *see also* Soil
        remediation
Pollution, 13; *see also* Toxic substance
        pollution
    forms of, 14–15
    from urban and agricultural runoff,
        14
Polychlorinated biphenyls (PCB), 4, 88;
        see also Toxic substance
    molecular structure of, 88
    properties, 89
Polycyclic aromatic hydrocarbons
        (PAHs), 87, 91–92; *see also* Toxic
        substance
Polyvinyl chloride (PVC), 242
Population models, 189–191; *see also*
        Ecological model
PPOM, *see* Particulate polycyclic
        organic matter (PPOM)
Precipitation and dissolution, 105; *see*
        also Chemical processes
Predicted environmental concentration
        (PEC), 161
PVC, *see* Polyvinyl chloride (PVC)
Pyrene, 92
Pyrethins, 90
Pyritic sulfur, 229
Pyrolysis, 242; *see also* Solid waste
        treatment

Q

QSAR (Quantitative structure activity
        relationships), 4

R

Radioactive contamination, 15; *see also*
        Pollution
Reach Reform, 279–280; *see also* Cleaner
        production
Reactions and equilibrium calculations,
        97, 153; *see also* Redox equilibria
    activity coefficients, 101–102

    adsorption, 110–112
    alkalinity, 117, 119
    buffer capacity, 116, 118
    carbonate solubility, 127–128
    chemical processes, 104–107
    complex formation, 131–133
    conditional constant, 133–136
    dissolved carbon dioxide, 119–124
    double logarithmic diagrams, 112–116
    equilibrium constant, 98–100, 102–103
    Henry's law, 108–110
    hydroxide solubility, 124–127
    molar fraction, 116
    pe as master variable, 142
    pe–pH diagram construction, 147–150
    redox conditions in natural waters,
        145–147
    redox potential and complex
        formation, 150–152
    relevant processes in aquatic
        environment, 142–145
    simultaneous reactions, 108
    solid phase stability, 129–130
    solubility of complexes, 128–129
Redox equations, 140; *see also* Redox
        equilibria
Redox equilibria, 139; *see also* Reactions
        and equilibrium calculations
    electron activity, 139
    example, 141–142
    free energy and pe, 140
    Nernst law, 140
    pe and redox potential, 139
    redox conditions in natural waters,
        145–147
    redox equations, 141
    redox potential and complex
        formation, 150–152
    standard redox potential, 140
Redox reactions, 106–107, 139; *see also*
        Chemical processes

S

Safety factors, 29
SDMs, *see* Structurally dynamic models
        (SDMs)
Simple DDT model, 190; *see also*
        Ecological model

Simultaneous reactions, 108; *see also* Reactions and equilibrium calculations

Sludge conditioning, 237–239; *see also* Solid waste treatment

Soaps, *see* Detergents

Software STELLA, 176, 177; *see also* Ecological model

Soil contamination, 15; *see also* Pollution

Soil remediation, 258; *see also* Ecological engineering; Ecotechnology
bioaccumulators, 260
extraction, 259
incineration, 259
in situ treatment of contaminated soil, 259–261
phytoremediation, 260–261
removal and treatment of contaminated soil, 258–259

Solid wastes, 234
classification of, 234
mass flow analysis, 235
recycling, 235
source, 236

Solid waste treatment, 236; *see also* Environmental technology
aerobic and anaerobic processes, 240
agricultural waste, 245
anaerobic digestion, 238–239
centrifugation, 238
comparison of methods, 243
composting, 240–241, 242
domestic garbage, 239–243
dumping grounds, 240
filter belts, 238
filter presses, 238
hospital wastes, 243–244
incineration, 241–242
industrial and mining waste, 244
microorganisms in, 241
pyrolysis, 242
sludge conditioning, 237–239
vacuum filtration, 238

Spheres, 97

Standard free energy, 98

Steady-state models of mass flows, 189; *see also* Ecological model

Structurally dynamic models (SDMs), 182; *see also* Ecological model

Sulfur cycle, 63–64

Sulfur dioxide; *see also* Air pollution abatement
pollution, 229
recovery, 230

Sulfuric acid, 228

Surfactants, 93; *see also* Toxic substance

Synthetic polymers, 93; *see also* Toxic substance

T

TDI, *see* Tolerable daily intake (TDI)

Teratogens, 34

Tetraethyl lead, 92; *see also* Toxic substance

Thermal pollution, 15; *see also* Pollution

Tolerable daily intake (TDI), 165

Toolbox environmental technology, 6; *see also* Integrated environmental management

Toxic substance, 67; *see also* Cadmium; Ecological model; Inorganic compound; Lead; Mercury; Organic compounds
adsorption, 31
biodegradability, 30
biodegradation rate, 31
detergents, 93, 94
distribution in environment, 27–32
ecotoxicological effects, 32–34
estimation methods for, 30
models, 187
organometallic compounds, 92–93
properties, 27, 28
safety factors, 29
synthetic polymers, 93
in trophic level, 189, 190
xenobiotics, 94

Toxic substance pollution, 13, 15; *see also* Pollution
bioaccumulation, 24
bioconcentration, 25
biomagnification, 25
distribution in time and space, 18–26
example, 23
fugacity, 20–23
GNP per capita, 17
harmful effect vs. dosage, 18

Toxic substance pollution (*Continued*)
  log exposure time vs. log LC_{50}, 25
  pollution sources, 13–18
  processes, 20
  quantification of, 16
  spatial and time scale for hazards and
      ecological hierarchy levels, 19
  sulfur effects on human health, 26
Triangle diagram, 133

U

UNCED, *see* United Nations
      Conference on Environment
      and Development (UNCED)
UNEP, *see* United Nations Environment
      Programme (UNEP)
United Nations Conference
      on Environment and
      Development (UNCED), 157
United Nations Environment
      Programme (UNEP), 270

V

Vacuum filtration, 238; *see also* Solid
      waste treatment
Visual pollution, 15; *see also* Pollution

W

Waste stabilization ponds (WSPs),
      255
Wastewater treatment, 216; *see
      also* Environmental
      technology
  cost of treatment, 221–222
  efficiency matrix of pollution
      parameters and,
      218–219
  hydraulic load reduction, 222
  methods, 217
  method selection, 222–223
  plants, 220–221
  precipitation by aluminum sulfate,
      224
  water pollution problems, 216
Water pollution, 15; *see also* Pollution
Work capacity, 211; *see also* Ecosystem
      services
Work energy, 208
WSPs, *see* Waste stabilization ponds
      (WSPs)

X

Xenobiotics, 94; *see also* Toxic substance

Printed and bound by CPI Group (UK) Ltd, Croydon, CR0 4YY

01/11/2024

01782637-0003